T0320424

Industrial Hygiene

Society has become more educated on the impacts on human health and environment, and there has been a noticeable decrease in the acceptance of this risk by workers and the public. However, to ensure a higher level of worker protection, a revised approach to industrial hygiene is needed focusing on risk-reduction. This second edition of *Industrial Hygiene* focuses on implementation of an industrial hygiene program, using a risk-based approach, in an operational environment.

Key elements of this book include balancing the art and science of industrial hygiene, risk-based industrial hygiene approaches, recognizing, evaluating, and controlling workplace hazards, medical monitoring for workers, emergency response, training, and evaluating industrial hygiene programs. Updated concepts on leadership, emergency response and monitoring, management of industrial hygiene projects, and risk communication are addressed in this edition. The book incorporates the ISO 45001 standard, "Occupational Health and Safety," into an industrial hygiene program, which further demonstrates a risk-based approach that is internationally recognized.

This is an ideal read for any student, researcher, or practitioner in the fields of occupational health and safety, industrial hygiene, risk management, and hazard control.

New case studies with proposed solutions underpin the learning, and the instructor resource pack comprising PowerPoint presentations, question bank, and resources to enhance learning are available in this edition for qualified textbook adoption.

Sustainable Improvements in Environment Safety and Health

Series Editor: Frances E. Alston, South Carolina, USA

Lean Implementation
Applications and Hidden Costs
Frances Alston

Safety Culture and High-Risk Environments
A Leadership Perspective
Cindy L. Caldwell

Industrial Hygiene
Improving Worker Health through an Operational Risk Approach
Frances Alston, Emily J. Millikin, and Willie Piispanen

The Legal Aspects of Industrial Hygiene and Safety
Kurt W. Dreger

Strategic Environmental Performance
Obtaining and Sustaining Compliance
Frances Alston and Brian Perkins

Occupational Exposures
Chemical Carcinogens and Mutagens
Frances Alston and Onwuka Okorie

Industrial Hygiene
Improving Worker Health through an Operational Risk Approach, Second Edition
Frances E. Alston and Emily J. Millikin

For more information about this series, please visit: https://www.crcpress.com/Sustainable-Improvements-in-Environment-Safety-and-Health/book-series/CRCSUSIMPENVSAF

Industrial Hygiene

Improving Worker Health through an Operational Risk Approach

Second Edition

Frances E. Alston and Emily J. Millikin

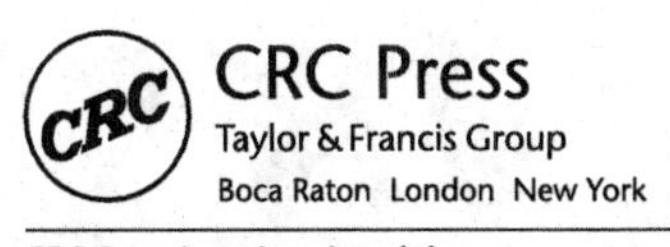

CRC Press is an imprint of the
Taylor & Francis Group, an informa business

Designed cover image: image credited to metamorworks [ShutterStock ID: 732758329]

Second edition published 2025
by CRC Press
2385 NW Executive Center Drive, Suite 320, Boca Raton FL 33431

and by CRC Press
4 Park Square, Milton Park, Abingdon, Oxon, OX14 4RN

CRC Press is an imprint of Taylor & Francis Group, LLC

First edition published by CRC Press 2018

ISBN: 9781032645889 (hbk)
ISBN: 9781032597645 (pbk)
ISBN: 9781032645902 (ebk)
ISBN: 9781032859422 (eBook+)

DOI: 10.1201/9781032645902

Typeset in Times
by Deanta Global Publishing Services, Chennai, India

Contents

Acronyms ... xi
Preface ... xiii
About the Authors ... xvi

Chapter 1 Occupational Safety and Health in the Workplace ... 1

1.1 Introduction ... 1
1.2 Multiple Facets of the Safety and Health Professional ... 3
1.3 Industrial Hygiene Program Tenets ... 6
1.4 Application and Implementation of Industrial Hygiene Program Elements ... 7
1.4.1 Program Management and Administration ... 7
1.4.2 Hazard Identification and Control Process ... 8
1.4.3 Occupational Health Management ... 8
1.4.4 Instrumentation and Calibration ... 9
1.4.5 Data Management, Records, and Reporting ... 9
1.4.6 Training and Qualification ... 10
1.4.7 Continuous Improvement ... 10
1.5 Industrial Hygiene Key Focus Area and Risk Reduction ... 11
Questions to Ponder for Learning ... 11

Chapter 2 Art and Science of Industrial Hygiene ... 13

2.1 Introduction ... 13
2.2 Art of Industrial Hygiene ... 13
2.2.1 Art of Hazard Recognition and Control ... 14
2.2.2 Art of an Occupational Exposure Monitoring Strategy ... 16
2.2.3 Art of the Occupational Health Program ... 17
2.2.4 Art of People Management ... 18
2.3 Science of Industrial Hygiene ... 19
2.3.1 Science of Hazard Recognition and Control ... 20
2.3.2 Science of an Occupational Exposure Monitoring Strategy ... 22
2.3.3 Science of the Occupational Health Program ... 22
2.3.4 Science of People Management ... 23
Questions to Ponder for Learning ... 24

Chapter 3 Industrial Hygiene Professional ... 25

3.1 Introduction ... 25
3.2 Role of the Industrial Hygiene Professional ... 26

3.3 Leadership and Organizational Structure 27
3.3.1 Flat Organizational Structure 28
3.3.2 Hierarchical Organizational Structure 28
3.4 Communication for Results 30
3.4.1 Technical and Non-Technical Workers and Colleagues 30
3.4.2 Risk/Exposure Assessment Data and Results 31
3.4.3 Relationship with the Workforce 31
3.4.4 Engagement on Work Planning Teams 32
3.5 Corporate Programmatic Support Role 32
3.6 Industrial Hygienist as an Expert Witness 33
3.7 Continuing Education and Professional Development 33
3.7.1 College and University Curricula 34
3.7.2 Retention of the Industrial Hygiene Professional 36
3.7.3 Industrial Hygiene Certification 37
3.7.4 Continuing Education 38
3.7.5 Job Rotation 38
3.7.6 Industrial Hygienist as a Generalist 39
3.8 Legal and Ethical Aspects of Industrial Hygiene 40
3.8.1 Professionalism 41
3.8.2 Accountability 41
3.8.3 Humility 41
3.8.4 Reliability 41
3.8.5 Trustworthiness 42
3.9 Management of Industrial Hygiene Projects 42
3.10 Emergency Response and Monitoring 43
Questions to Ponder for Learning 43
References 44

Chapter 4 Strategies for Exposure Monitoring and Instrumentation 45

4.1 Introduction 45
4.2 Regulatory Aspects of Industrial Hygiene Monitoring 47
4.3 Quantitative and Qualitative Exposure and Risk Assessment 48
4.3.1 Quantitative Exposure and Risk Assessment 48
4.3.2 Qualitative Exposure and Risk Assessment 49
4.4 Process Flow of Exposure Assessment 50
4.4.1 Defining the Scope of Work 52
4.4.2 Developing a Monitoring Plan 53
4.4.3 Implementing the Characterization and Monitoring Plan 55
4.4.4 Similar Exposure Groups 56
4.4.5 Occupational Exposure Control Banding 57
4.4.6 Evaluating Data and Characterizing Exposures 58

4.4.7 Develop Controls 61
4.4.8 Document Results 62
4.4.9 Communicate Data Results 63
4.4.10 Develop a Re-Evaluation Plan 64
4.5 Occupational Safety and Health Characterization and Monitoring Equipment 64
4.5.1 Diffusion Detector Tubes 65
4.5.2 Vapor Monitor Badges 66
4.5.3 Personal Air Sampling Pumps 67
4.5.4 Handheld Electronic Monitors 67
4.5.5 Fixed Air Monitors 68
4.6 Case Studies To Facilitate Thoughtful Learning 68
4.6.1 The Presence of an Intermittent Odor 68
4.6.2 The Presence of an Intermittent Odor Lessons Learned 69
4.6.3 "I Have Been Sick for the Past 6 Months" 69
4.6.4 "I Have Been Sick for the Past 6 Months": Lessons Learned 70
Questions to Ponder for Learning 71
References 71

Chapter 5 Risk-Based Industrial Hygiene 72

5.1 Introduction 72
5.2 Importance of Risk Assessments and a Risk-Based Approach to Hazard Management 73
5.3 Identifying and Controlling Workplace Risks 73
5.4 Addressing Industrial Hygiene Risks in the Workplace 74
5.4.1 Industrial Hygiene Risk Assessment 75
5.5 Risk Ranking 77
5.6 Integration of a Risk-Based Consensus Standard Into Industrial Hygiene 78
5.6.1 Industrial Hygiene Program Management and Administration 79
5.6.2 Hazard Identification and Control Process 80
5.6.3 Occupational Health Management 81
5.6.4 Instrumentation and Calibration 81
5.6.5 Data Management, Records, and Reporting 81
5.6.6 Training and Qualification 82
5.6.7 Continuous Improvement 82
5.7 Risk Communication 82
5.8 Risk Acceptance 84
Questions to Ponder For Learning 84
Reference 85

Chapter 6 Recognizing, Evaluating, and Controlling Workplace Hazards 86

6.1 Introduction 86
6.2 Historical Chemical and Industrial Hazards 89
6.3 Chemical, Physical, Biological, and Industrial Hazards of the Past Decade 92
6.4 Workplace Hazard Inventories 94
6.4.1 Task Hazard Inventory 94
6.4.2 Facility Hazard Inventory 95
6.5 Injury and Illness Logs and Inspection Trending 95
6.6 Chemical Inventories: Use, Storage, and Disposal Records ... 97
6.7 Biological and Radiological Hazards Considerations 100
6.8 Regulatory Inspections and Violations 100
6.9 Hazard Control and Work Execution 101
6.9.1 Hazard Elimination 102
6.9.2 Product Substitution 102
6.9.3 Engineering Controls 102
6.9.4 Work Practices and Administrative Controls 102
6.9.5 Personal Protective Equipment 102
6.10 Integration of Hazard Recognition and Controls and Work Control Processes 102
6.10.1 Planning Work 103
6.10.2 Authorizing Work 103
6.10.3 Work Execution 105
6.10.4 Project Closeout 105
6.11 Management by Walk-Around 107
6.11.1 Preparation 107
6.11.2 Communication 108
6.11.3 Persistence to Connect 108
6.12 Safety Through Design: Designing Hazards Out of the Process 108
6.13 Employee Engagement and Involvement 109
Questions to Ponder for Learning 111
References 111

Chapter 7 Medical Monitoring and Surveillance of the Worker 113

7.1 Introduction 113
7.2 Medical Monitoring and Surveillance Program 114
7.3 Establishment of Company Policies, Protocols, and Procedures 115
7.4 Scheduling and Tracking of Physicals 117
7.5 Interfacing with Medical Professionals 120
7.6 Notification of Test Results 121
7.7 Analysis of Occupational Health Data 121

7.8 Medical Monitoring Records and Reporting 122
7.9 Case Study 1: Chromium IV Exposure 123
7.10 Case Study 2: Beryllium Exposure 124
Questions to Ponder for Learning........ 125
Reference........ 125

Chapter 8 Workforce Training on Hazard Recognition and Control........ 126

8.1 Introduction 126
8.2 Why Provide Workplace Training? 127
8.3 Developing an Effective Training Strategy 128
8.4 Hazard Recognition, Evaluation, and Control Training........ 129
8.4.1 Hazard Anticipation 130
8.4.2 Hazard Recognition and Identification........ 131
8.4.3 Evaluate Hazards........ 132
8.4.4 Controlling Hazards 133
8.5 Trainer Knowledge and Qualification 134
8.6 Training Effectiveness Evaluation........ 135
8.7 Other Training Methods and Tools 136
8.7.1 Process or Tool Mock-Up........ 136
8.7.2 Peer-to-Peer Training 136
8.7.3 Training on the Job Site........ 137
8.7.4 Hazard Identification Checklist........ 137
8.7.5 The Use of Case Studies to Enforce Learning........ 137
8.7.6 Microlearning........ 138
Questions to Ponder for Learning........ 143
References 144

Chapter 9 Industrial Hygiene and Emergency Response........ 145

9.1 Introduction 145
9.1.1 Hurricane Katrina – United States........ 145
9.1.2 La Porte, Texas Chemical Plant – United States........ 146
9.1.3 September 11, 2001 – United States........ 146
9.1.4 Collapse of the Rana Plaza........ 147
9.1.5 Glasgow Explosion – Scotland........ 147
9.2 Approaches to Emergency Response........ 148
9.3 Initiating Event 150
9.4 Event Notification 151
9.5 Event Response........ 154
9.5.1 Risk Prioritization 154
9.5.2 Resources and Equipment........ 155
9.5.3 Logistical Support........ 156
9.5.4 Event Response Communication........ 157
9.6 Event or Site Transition 158

9.7 Lessons Learned from 9/11 ... 159
9.7.1 Event Notification ... 159
9.7.2 Event Response ... 160
9.7.3 Site Transition ... 161
Questions to Ponder for Learning ... 162
References ... 162

Chapter 10 Evaluating the Industrial Hygiene Program ... 163

10.1 Introduction ... 163
10.2 Identifying the Program and Process to Assess ... 164
10.3 Identifying Key Attributes to Assess ... 165
10.4 Developing the Assessment Plan and Lines of Inquiry ... 166
10.5 Performing the Assessment ... 168
10.6 Data Analysis ... 170
10.6.1 Data Organization ... 170
10.6.2 Analysis Method ... 171
10.6.3 Risk Management of Identified Hazards ... 172
10.6.4 Relationship of Data to Existing Programs and Processes ... 173
10.6.5 Data Storage and Management ... 173
10.7 Document Assessment Results ... 175
Questions to Ponder for Learning ... 176

Chapter 11 Continuous Improvement ... 177

11.1 Introduction ... 177
11.2 Continuous Improvement Process ... 178
11.3 Establishing a Performance Baseline ... 180
11.4 Identify Areas for Improvement ... 180
11.5 Industrial Hygiene Continuous Improvement Plan ... 182
11.5.1 Goals and Objectives ... 183
11.5.2 Corrective Actions and Improvement Initiatives ... 183
11.6 Performance Metrics and Monitoring ... 185
11.7 Case Study: Tungsten Tools ... 188
Questions to Ponder for Learning ... 189

Index ... 191

Acronyms

9/11	September 11, 2001
ABET	Accreditation Board for Engineering and Technology
ABIH	American Board of Industrial Hygiene
ADA	Americans with Disability Act
AIHA	American Industrial Hygiene Association
ALARA	As Low As Reasonably Achievable
AOC	Administrative Order on Consent
ASTM	American Society for Testing and Materials
BEAR	Biological Effects of Atomic Radiation
BEIs	Biological Exposure Indices
CERCLA	Comprehensive Environmental Response Compensation and Liability Act
CLP	Classification, labeling, and packaging
CSB	U.S. Chemical Safety and Hazard Investigation Review Board
EEOICP	Energy Employees Occupational Illness Compensation Program Act
EPA	U.S. Environmental Protection Agency
EPCRA	Emergency Planning and Community Right-to-Know Act
ES&H	Environmental safety and health
EU	European Union
GHS	Global harmonization system
HSE	Health and Safety Executive
IARC	International Agency for Research on Cancer
ISO	International Organization for Standardization
LNT	Linear no-threshold model
LT	Linear threshold model
MSDS	Material safety data sheet
NIOSH	National Institute of Occupational Safety and Health
PDCA	Plan-Do-Check-Act
OEB	Occupational exposure banding
OEL	Occupational exposure limit
OECD	Organization for Economic Co-Operation and Development
OH&S	Occupational Health and Safety
OSHA	Occupational Safety and Health Administration
PCBs	Polychlorinated biphenyls
PPE	Personal protective equipment
REL	Recommended exposure limit
RCRA	Resource, Conservation, and Recovery Act
RIDDOR	Reporting of Injuries, Diseases, and Dangerous Occurred Regulations
SARA	Superfund Amendments and Reauthorization Act
SEG	Similar exposure group
SMART	Specific, measurable, achievable, relevant, and timebound

TLV	Threshold limit value
TVA	Tennessee valley authority
TWA	Time-Weighted Average
UK	United Kingdom
US	United States
WTC	World Trade Center

Preface

Over the past century, the discipline of industrial hygiene has focused on protecting employees from hazards in the work environment. Since the passage of the Williams-Steiger Act of 1970, the industrial hygienist has played a key role in the prevention of injuries and illnesses in the workplace. Anticipating, recognizing, and controlling hazards are the favored approaches used by industrial hygienists in protecting the worker from hazards in the workplace. Consequently, the industrial hygienist must have the knowledge and skills to analyze and understand risks, along with understanding how to effectively mitigate and control hazards. In many cases, hazards in the workplace exhibit more than one risk factor.

The hazard identification and control process is based on effectively understanding the inter-relationship between the three process elements of hazard identification, analysis, and control. Physical hazards are the most commonly encountered hazards in the workplace; however, chemical, biological, or radiological hazards may also be present either individually or in combination with other hazards.

Traditional industrial hygiene has been focused on collecting and analyzing data in order to identify and characterize a variety of workplace hazards. Although today's industrial hygienists are taught various methods and techniques on how to recognize and manage occupational hazards, it is generally a knowledge- and skill-based approach to hazard recognition and control that is used to help render workplaces free of hazards. Each workplace and situation is unique and the industrial hygienist must develop and adapt scientifically based methods to understand and control hazards. However, when the industrial hygienist works in the field, and as they interact with the workers, they must understand how to communicate effectively with workers and management when gathering, evaluating, and communicating sampling and monitoring results, and conclusions. Both the industrial hygiene program and the implementation of the program in the field should be evaluated and improved where feasible, to ensure compliance with the most current requirements, standards, and safety practices and to ensure improvement is in alignment with a risk-based philosophy and approach. Often, goals and steps needed to complete improvement initiatives are written into a continuous improvement plan for industrial hygiene. The value of such improvements can be realized and equated into additional methods and avenues for risk-reduction.

Industrial hygiene is not an exact science and is influenced by people, professional judgment, science, and mathematics; it is a profession driven by both art and science. Often, the industrial hygiene discipline is viewed as a specialized function and the industrial hygienist is frequently viewed as a consultant although they are an integral functioning part of the team. The practicing industrial hygienist is tasked with assisting management in achieving a reduction in workplace injuries and illnesses in a corporate environment that is increasingly looking for more cost-effective and efficient methods of conducting business. At times, these efficiencies may target a reduction in the amount of labor needed to perform a job, less expensive materials to use in their manufacturing processes, and more efficient methods to manufacture a product. The industrial hygienist is faced with trying to assist in the achievement of these increased performance goals, with traditional industrial hygiene practices and principles that may, or may not, result in an increase in worker protection and a decrease in health risks posed by workplace hazards.

Over the past 40 years, the industrial hygiene profession has significantly grown and is expected to continue to grow as workplaces continue to evolve in the production and usage of hazardous materials, consistent with a shift in public opinion regarding the acceptance of the health risk from activities performed at work and at home. As time progresses, industries are being regulated to not only minimize the health impacts to the workforce but also decrease the likelihood of impacting the environment.

Society has become more educated on the impacts to human health and environment, and there has been a noticeable decrease in the acceptance of this risk by workers and the public. Corporations and institutions are faced with managing the impacts, both financially and from a public perception perspective, resulting from this decreased acceptance of workplace injuries and illnesses. In particular, a vast amount of information on safety and health is readily available on the internet, which can add additional pressures and further complicate the way companies and institutions communicate how they are effectively managing safety in the workplace. The accepted standard of performance for industrial hygiene has grown beyond compliance, but now also focuses on improving existing processes and creating a workplace *free* of injury and illness.

A revised approach to industrial hygiene is needed which is focused, and tailored, on risk-reduction to realize an increased level of worker protection. The industrial hygienist will continue to evolve and must grow using a risk-based approach in order to better perform their job and improve the protection of the worker. This book focuses on the implementation of industrial hygiene, using a risk-based approach, in an operational environment. The second edition integrates the ISO 45001 standard into a risk-based industrial hygiene program. Key elements of this book include:

- Balancing the Art and Science in Industrial Hygiene
- The Industrial Hygiene Professional
- Risk-Based Industrial Hygiene
- Recognizing, Evaluating, and Controlling Workplace Hazards
- Medical Monitoring for Workers
- Emergency Response
- Training
- Evaluating the Industrial Hygiene Program
- Continuous Improvement

The approaches and methods described in this book are designed to assist the industrial hygienist in managing workplace risks, including risks associated with anticipation, recognition, evaluation, and hazard control processes. A focus on risk reduction or prevention will move a company closer to the goal of an injury-free workplace, while improving methods and techniques used by the industrial hygienist for implementing an effective worker protection program.

About the Authors

Frances E. Alston has more than 35 years of leadership experience in high hazards, diverse work cultures. She is an instructor at a major university, a senior manager in an international company, and the Past President (2018) of the American Society for Engineering Management. Dr. Alston is also a fellow of the American Society for Engineering Management (ASEM) and has built a solid career leading the development and management of Environment, Safety, Health, and Quality (ESH&Q) programs in challenging cultural environments. She has a Ph.D. in Industrial and System Engineering and an MSE degree in Engineering Management, both from the University of Alabama. She earned a master's degree in Hazardous and Waste Materials Management/Environmental Engineering from Southern Methodist University and a Bachelor of Science degree in Industrial Hygiene and Safety/Chemistry from Saint Augustine's University. She holds certifications as a Certified Hazardous Materials Manager (CHMM) and a Certified Professional Engineering Manager (CPEM).

Emily J. Millikin has over 39 years of leadership experience in regulatory, environmental, radiation protection, and safety and health at Department of Energy (DOE) and Department of Defense (DOD) chemical and radiological operations and remediation. She has held various executive leadership positions in different operational environments and is a proven leader in achieving excellence in both program and field execution of safety and health, radiation protection, quality assurance, environmental, industrial hygiene, safety culture, and voluntary protection programs. Ms. Millikin earned a B.S. in Environmental Health with majors in both Industrial Hygiene and Health Physics from Purdue University. She is a Certified Industrial Hygienist (CIH) and Certified Safety Professional (CSP) and serves as an Advisory Board Member to the DOE National Supplementary Screening Program. Ms. Millikin currently provides consulting services to the DOE at the Hanford Site, Washington.

1 Occupational Safety and Health in the Workplace

1.1 INTRODUCTION

Long before there were safety and health and industrial hygiene professionals, with varying degrees and certifications, the field of occupational safety and health existed. Within the United States most safety professionals recognize that one of the earliest forms of legislation associated with the safety profession was the Massachusetts Factory Act of 1877. The Massachusetts Factory Act created an inspection process for factory equipment, such as machinery, elevators, and workplace fire exits. Over the next century, additional programs and legislation were introduced to improve environmental and health conditions in the workplace. The practice of safety and health in other countries has also evolved in other countries across the globe. The industrial revolution occurred in the United Kingdom (UK) between 1760 and 1830. During this time frame rapid changes in urbanization associated with dirty and unpleasant living conditions and epidemics of infectious diseases occurred. As a result, during the succeeding decades many health and safety legislation was enacted. Working and residential conditions and their health impact on adults and children were documented in the 1830s by Charles Turner Thakrah, known as the UK father of occupational medicine. In 1974 the Health and Safety and Work Act was introduced that defines the authority and enforcement of the regulation and outlines the responsibility of employees and employers. The Health and Safety Executive (HSE) was founded in January 1975 having responsibility for enforcing the act. The Health and Safety Executive can be viewed as being equivalent to and serving the same purpose as the Occupational Safety and Health Administration in the United States.

In 1913, the United States Congress created the Department of Labor, with a focus on promoting the health and welfare of workers, improving workplace conditions, and ensuring workplace benefits. In 1935, the Social Security Act was enacted, which allowed the funding of state health departments and industrial health programs. The culmination of these efforts resulted in President Nixon approving the Williams–Steiger Act of 1970, often referred to as the Occupational Safety and Health Act (OSHA), which was intended to ensure that workers had safe and healthy working conditions.

The United States Congress determined that personal injuries and illnesses, originating from workplace conditions, imposed a substantial burden on, and a hindrance to, interstate commerce (across states). This burden was viewed in terms of lost production, wage loss, medical expenses, and disability compensation payments. Congress declared its purpose and policy, through the exercise of its powers,

DOI: 10.1201/9781032645902-1

to regulate commerce among the several states and with foreign nations, and to provide, for the general welfare of every working man and woman in the nation, safe and healthful working conditions and to preserve our human resources. This was a profound piece of legislation because the governing body of the United States recognized that harming their people, while doing business, was not acceptable and could ultimately lead to damaging the fundamental statehood of the country. As a result, Congress directed the focus of the Occupational Safety and Health Act to:

- Encourage employers and employees in their efforts to reduce the number of occupational safety and health hazards at their places of employment, and to stimulate employers and employees to institute new and improved programs for providing safe and healthy working conditions.
- Provide employers and employees separate but dependent responsibilities and rights with respect to achieving safe and healthy working conditions.
- Authorize the Secretary of Labor to establish mandatory occupational safety and health standards applicable to businesses affecting interstate commerce, and to create an Occupational Safety and Health Review Commission for carrying out adjudicatory functions under the act.
- Build on advances already made through employer and employee initiatives to provide safe and healthful working conditions.
- Provide for research in the field of occupational safety and health, including psychological factors involved, and the development of innovative methods, techniques, and approaches for dealing with occupational safety and health problems.
- Explore ways to discover latent diseases, establishing causal connections between diseases and work in environmental conditions, and conducting other research related to health problems, in recognition of the fact that occupational health standards present problems often different from those involved in occupational safety.
- Provide medical criteria that would ensure, in so far as practicable, that no employee will suffer diminished health, functional capacity, or life expectancy as a result of his work experience.
- Provide training programs to increase the number and competence of personnel engaged in the field of occupational safety and health.
- Provide for the development and promulgation of occupational safety and health standards.
- Provide an effective enforcement program that would include a prohibition against giving advance notice of any inspection, and sanctions for any individual violating this prohibition.
- Encourage the states to assume the fullest responsibility for the administration and enforcement of their occupational safety and health laws by providing grants to the states to assist in identifying their needs and responsibilities in the area of occupational safety and health.
- Develop plans in accordance with the provisions of this act.
- Improve the administration and enforcement of state occupational safety and health laws.

- Conduct experimental and demonstration projects in connection therewith.
- Provide appropriate reporting procedures with respect to occupational safety and health that would help achieve the objectives of the act and accurately describe the nature of the occupational safety and health problem.
- Encourage joint labor management efforts to reduce injuries and disease arising out of employment.

Over the past four decades, the United States Occupational Safety and Health Administration (OSHA), individual state agencies, industries and academics, and every workplace have seen a significant decrease in the number of work-related fatalities. Worker fatalities have decreased from approximately 38 worker deaths per day to about 13 worker deaths per day in 2015 (OSHA 2017). The occupational safety and health professional may or may not be an industrial hygienist, but the industrial hygiene discipline has traditionally been viewed as a specialized subset of the safety and health profession. The profession is typically focused on implementing those goals of the Occupational Safety and Health Act that involve determining the relationship between diseases and work in environmental conditions; implementing processes and procedures for identifying, analyzing, and controlling workplace hazards; monitoring to ensure compliance with regulatory standards; and promoting the performance of work in a safe and compliant manner.

Since the passage of the Occupational Safety and Health Act, the industrial hygiene profession has significantly grown, and is expected to continue to grow, consistent with a shift in public opinion regarding the acceptance of health risk from activities performed at work and at home. As society has become more educated on the impacts of chemicals on human health and the environment, there has been a decrease in the acceptance of this risk by workers and the public. The accepted standard of performance has grown beyond compliance, but now also focuses on improving existing processes and creating a workplace *free* of injury and illness.

No matter what the industry is, the workforce has greater access to information related to health risks from chemicals and other hazards associated with the work environment. In addition, more industries are being regulated to not only minimize the health impacts to the workforce but also decrease the risks of impacting the environment. As we progress into the twenty-first century, decreasing risks to human health and the environment is projected to continue to grow and evolve, and the role of the industrial hygienist must also evolve to address the growing needs of society, companies, industries, and the workforce.

The evolution of safety and health continues to evolve, and regulations being enacted to further improve health and safety of workers.

1.2 MULTIPLE FACETS OF THE SAFETY AND HEALTH PROFESSIONAL

Human beings have a need to feel safe in all aspects of their life. Maslow's hierarchy of needs identifies safety as a fundamental need of human life (Maslow 1954). Maslow's theory consists of five needs, or stages, which motivate humans to perform work and improve their life. Simply stated, the most basic needs must be met before

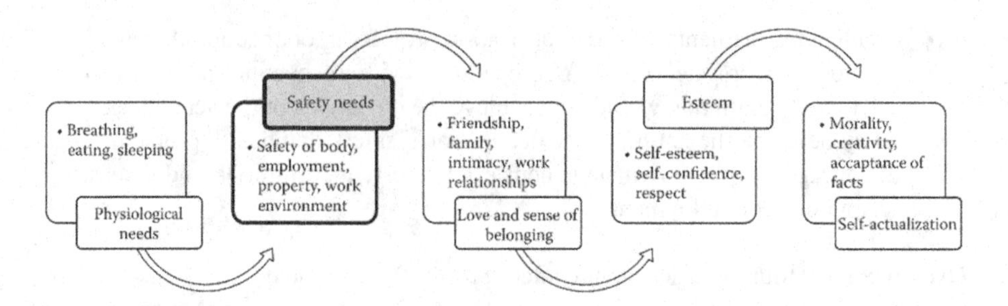

FIGURE 1.1 Maslow's hierarchy of needs process.

humans will be motivated to achieve the next level of improvement to their physical and mental being. Each need, or stage, must be met before a person can progress to the next stage. Figure 1.1 is a depiction of Maslow's theory.

The first tier of the hierarchy addresses the most basic need of people, and workers, which is physical well-being. A person cannot or will not move forward, or be driven to move forward, in achieving goals and objectives and improving their mental state if fundamental needs, such as food, water, and sleep, are not met. A human being may not be able to work on a consistent basis if his or her psychological and physiological needs are not met. One needs to look no further than the disaster in Flint, Michigan (Genesee County Board of Commissioners 2015), and the inadvertent cause of lead contamination in residential water. The ability of people to achieve the most fundamental step of Maslow's hierarchy of needs was impacted and were forced to find alternative means to address basic hygiene needs. Consequently, these residents must live, for many years, with the ramifications that their most basic needs as a human being were taken from their lives and families. It will be very difficult for some residents, and their families, to move forward and feel safe when drinking water from their home.

The need for humans to feel safe was recognized in the writings of Maslow in 1954. In order for workers to progress through the hierarchy of needs and become all that they can become, the safety needs, the second step in the process, must be fulfilled. Workplace safety enhances the ability for workers to become active, productive, and contributing team members that are capable of solving problems and complete work tasks. Therefore, workplace safety is an integral component of a leader's ability to motivate and influence workers.

Once a worker's physiological needs are met, the second tier of the hierarchy of needs focuses on people's safety and the security of their surroundings, whether it is to preserve meeting the first hierarchy of need or to improve their life. Not only is safety recognized as being important to the well-being of the human, but it is also important in other aspects of a human's life. For example, without achieving a safe work environment, workers cannot become motivated to develop a sense of belonging, as part of a team, without the fundamental premise of a safe work environment. The industrial hygiene professional is one facet of a company or institution that assists in achieving and maintaining a safe work environment that meets the needs of, and protects the worker.

Over the past four decades, the role of the industrial hygiene professional has evolved. Below are some examples of the roles and responsibilities of the industrial hygienist in today's modern work environment.

- Worker advocate: The industrial hygiene professional must have a good relationship with the workforce and be a good listener; he or she is the first line of defense for a worker to communicate improvements or problems in performing a work task.
- Company advocate: The industrial hygiene professional is employed by the company as part of the company's responsibility for providing a workplace free from hazards. The industrial hygiene professional always bears that responsibility and is certainly viewed as an expert or doer for ensuring that conditions in the workplace are free from hazards.
- Advocate for compliance with regulatory, contract, and company requirements: The industrial hygiene professional has the primary responsibility for monitoring and ensuring that companies and industries are compliant with regulatory limits for non-radiological hazards.
- Facilitator and sponsor for safe work activities: Often, the industrial hygiene professional works to facilitate communications between management and the workforce with respect to safety initiatives and concerns.

A good industrial hygienist establishes safety and health goals and standards that not only comply with regulatory requirements but also surpass those standards to promote a robust and improved safety program. The industrial hygienist is in a position to be a leader in the promotion of improving standards for workers, for the company, and across the industry. It is also worth noting that as the need for industrial hygienist professionals grows, so too does the professional liability associated with the position.

Within the past 10 years, a trend has started to emerge with respect to the level of accountability and liability associated with the industrial hygiene position. For many years, the industrial hygiene profession was thought of as another work position that did not require additional training or skills – as opposed to those recognized as being required for engineers, operations and maintenance (skill-based specialties) leaders, or the performance of research and development activities. Since the passage of the Occupational Safety and Health Act; Resource, Conservation, and Recovery Act (RCRA); and Comprehensive Environmental Response, Compensation, and Liability Act (CERCLA), the liability of the industrial hygiene position has significantly increased.

Safety and health practices associated with standard industrial hazards are generally associated with a well-defined source and waste stream, and that standard may not be adequate to protect workers who are performing remediation work which often lacks well-defined contaminants and contamination. Because many of these waste sites originated when safety and health requirements did not exist, performing work at these types of locations present challenges to the industrial hygiene professional because of the need to protect workers from additive and synergistic health effects from multiple contaminants. It is up to the industrial hygiene professional to clearly understand the professional risk, as well as human health risks, posed by work activities and the overall work environment. As such, the skill set of the industrial hygienist must address this growing need for a technically defensible, risk-based approach to industrial hygiene.

1.3 INDUSTRIAL HYGIENE PROGRAM TENETS

There are several industrial hygiene books available, which discuss how to perform and implement industrial hygiene work activities. However, what is often not communicated, and often misunderstood, is the purpose and focus of why these activities are being performed. As an industrial hygiene professional, the job activities will vary depending on the company or industry, but fundamentally, there are common tenets, or beliefs, held by the industrial hygiene professional. Table 1.1 lists a few of these tenets that are common across companies and industries. Several of them are discussed in further detail throughout this book. These tenets are based on attributes needed to meet and exceed the intent of meeting both regulatory and consensus-based safety and health standards. Additionally, these tenets further support the original mission of the United States Department of Labor and Congress, with a focus on promoting the health and welfare of workers – which was recognized as one of the most valuable resources of a company or business, along with interstate commerce. The industrial hygienist is integral to achieving a reduction in the link between poor workplace conditions and occupational diseases and illnesses, as demonstrated over the past 200 years.

Table 1.1 lists only a few tenets that are integral to an industrial hygienist, but several of them are integral to establishing and building relationships with the workforce and building a safety and health program that is focused on the prevention of injuries and illnesses versus being reactive to safety incidents.

TABLE 1.1
Examples of Industrial Hygiene Program Tenets

Perform work with integrity and in a quality manner.	Ensure that workers are aware of hazards posed when performing work.
Trust, and the establishment of trust, is a key attribute in the performance of the safety and health and industrial hygiene discipline.	Facilitate to ensure that joint ownership of the safety program between management and the workforce exists.
Verification and validation of compliance with state and federal regulations are required to ensure that work is being compliantly conducted.	Promote workplace and job satisfaction working as a team.
Strive to improve working conditions to minimize health risks posed to workers.	Work products are produced on schedule and under budget.
Build and implement a program that is protective of the workforce.	Workplace injuries and illnesses are reduced.

1.4 APPLICATION AND IMPLEMENTATION OF INDUSTRIAL HYGIENE PROGRAM ELEMENTS

There are any number of ways an industrial hygiene program can be established; however, all industrial hygiene programs include processes, policies, and procedures that establish and implement the industrial hygiene program elements. Figure 1.2 depicts the common elements of an industrial hygiene program. A general summary of each of these elements is presented below.

The industrial hygiene function may be organized as a subset of the overall safety and health program, or may be managed as its own organization or group depending on the level of risk posed by work activities to the worker (e.g., if the company mission deals with chemicals or other toxicological hazards, then additional organizational focus on diseases or illnesses may be needed in managing work activities).

1.4.1 Program Management and Administration

This functional element establishes and manages policies, plans, and procedures that have been developed to define the strategy, approach, and documentation used to implement the industrial hygiene program. The organizational structure and budget are managed within this functional element, along with ensuring the identification of all applicable regulations and consensus standard guidelines or requirements (i.e., ISO 45001), adequate flow-down of requirements, the definition of work to be performed by industrial hygienists, funding, any schedules needed to better manage work activities, and mechanisms that verify the submittal of all regulatory required reports, such as the OSHA 300 log or a risk management log.

This functional element is key in the implementation of the industrial hygiene program because foundational processes and procedures, which define risk objectives and criteria by which the industrial hygienists manage and perform work, such

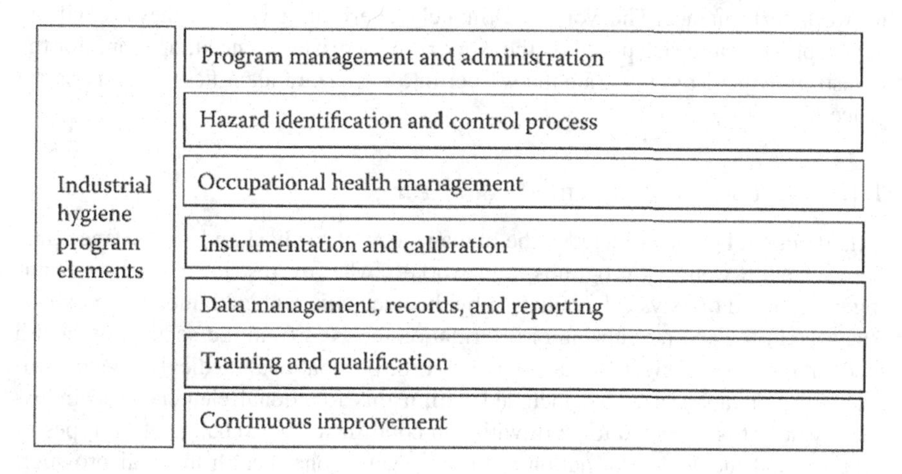

FIGURE 1.2 Common industrial hygiene program elements.

as the exposure assessment process, hearing conversation program, and emergency response protocols, are established. The approach, integration, and management of risk will also be defined and integrated across all industrial hygiene program elements based on policies and procedures established.

1.4.2 Hazard Identification and Control Process

This functional element encompasses all work activities necessary to identify and control hazards associated with routine and nonroutine work (e.g., responding to an equipment breakdown). Evaluation of the individual and collective work scope and work steps, implementation of the exposure assessment process, identification and application of hazard controls, and documentation of the hazard evaluation process are performed within this functional element. Also included is the application of the hazard control hierarchy, which includes product or equipment substitution, engineering controls, administrative controls, and personal protective equipment. It is worth noting that the hazard identification and control process is not just limited to work being conducted in the context of producing a product, but also includes design review activities (commonly referred to as safety through design).

The hazard identification and control process defines a large part of the industrial hygiene discipline because processes and procedures associated with the hazard identification and control process can be used to significantly reduce the health risk posed by a work activity or toxicity of contaminants. Included in this functional element is the implementation of processes that define the methods by which industrial hygienists will evaluate and determine the health risk associated with the work and necessary controls to mitigate the hazards, along with how to perform the collection of sampling and monitoring information. Ultimately, the health risk posed by conditions that remain after applying the hazard identification and control process will then be considered by workers when they are determining whether to accept the risk while performing the job as defined or provide feedback as to alternative methods for work performance. The workers ultimately determining whether they are willing to accept the health risk posed by the work activity drives the need, up front, for the industrial hygienist to integrate the worker into the hazard identification and control process.

1.4.3 Occupational Health Management

This functional element includes the management of medical and surveillance programs, health wellness programs, and workers who are restricted from performing work based on physical evaluation by the occupational health medical provider. The scheduling and monitoring of information received from the occupation health medical provider, along with management of the occupational medical provider contract (if applicable), are also included within this functional element. The industrial hygienist is expected to work with and coordinate the exchange of all types of occupational medical information with the occupational health medical provider; the industrial hygienist is the company's primary representative in ensuring that workplace conditions, such as air and surface concentrations of contaminants, are communicated to the medical provider in determining whether a workplace injury or illness has occurred. This information is also frequently provided to the workers' compensation group. It is the occupational health medical provider who is key to the industrial hygienist in establishing the health risk and link between occupational exposure and physiological responses and illnesses. This functional element is one that is expected to continue to grow and evolve over the next decade because of technological advances in medicine and the ease by which the general working population has access to occupational disease information through the internet.

1.4.4 Instrumentation and Calibration

This functional element focuses on the establishment of an instrumentation program, which could include direct-reading instruments, sampling pumps, and stationary monitoring devices. Included within this element are the standards used for calibration and maintenance, and the periodicity for performing equipment checks, along with ensuring that quality assurance requirements are being met, which may include additional requirements imposed on the industrial hygiene function if performing work in a high-hazard work environment (e.g., nuclear or chemical weapons programs). A key aspect of the instrumentation and calibration program is understanding quality assurance requirements that may be invoked because of the industry in which the program is implemented. For example, an instrumentation and calibration program may be required to meet the elements of the Nuclear Quality Assurance requirements if it is implemented in a company or industry that has adopted the Nuclear Quality Assurance-1 standard for radiological work.

1.4.5 Data Management, Records, and Reporting

This functional element, although it fundamentally seems to be simple, includes all the data management, record keeping, and reporting of information to individuals, the company, and regulatory bodies. Data collection associated with sampling and monitoring, along with information related to ensuring compliance with regulatory requirements, is managed within this element. The amount of labor needed to perform adequate analysis of data has exponentially increased with today's use of computers, whether the data management is associated with website maintenance (since most companies have their own safety and health or industrial hygiene web page) or instrumentation that continuously transmits via Bluetooth, and the ability to be able to organize, evaluate, and use this information real-time has become an issue with many companies because the resources needed to perform such tasks is not usually planned as part of work. In particular, if the company routinely performs sample collection, or the operational mission uses industrial hygiene monitors for process monitoring purposes, the amount of time and money needed to organize, evaluate, and develop conclusions continues to increase because of the cost in labor needed to process the information. The need to plan and allocate additional time and resources to analyze and determine how data can be effectively utilized by companies is quickly becoming an emergent issue within the industrial hygiene profession. As innovations within artificial intelligence become realized, perhaps the ability to quickly process monitoring, in particular real-time data, may also become a greater reality.

1.4.6 Training and Qualification

The industrial hygiene functional element for training and qualification may be organized by two types of training: (1) training received as a professional and (2) training provided by the industrial hygienist to others in need of specialized training (e.g., confined space, fall protection, and electrical safety). Training received as an industrial hygienist may include the following:

- Basic training received as a new hire: Most companies have some type of mandatory training required as part of the hiring onboarding process. Generally, this training may consist of company policies and procedures related to human resource topics, security, safety and health policies, and return to work.
- Job-specific training: Professional training tailored to company-specific hazards. For example, industrial hygienists that support industries who manage highly hazardous chemicals may be required to attend emergency response-specific training.
- Training received by the industrial hygienist as part of the industrial hygiene qualification and continuing education process: If training is provided as part of a qualification process, as an industrial hygienist, the training may be organized as company-defined industrial hygiene professional, company-defined industrial hygiene supervisor, or company-defined industrial hygiene qualified subject matter expert (e.g., instrumentation).

In many companies, the industrial hygienist teaches general safety and health classes, along with classes that may be skill specific, and industrial hygiene topics such as confined space or respiratory protection. Health risks to the worker, associated with the topic that is being instructed, should be effectively communicated to further emphasize the importance of correctly performing the work activity.

1.4.7 Continuous Improvement

Industrial hygienists should always seek to improve both the performance and the quality of their work. Continuous improvement can be demonstrated across all industrial hygiene functional elements. In particular, improvements in the hazard identification and control process can lead to a reduction in the number of unsafe work practices. Continuous improvement can occur on both an individual performance basis and across the company. Many companies have safety initiatives that are tailored to focus on reducing injuries associated with a specific work function, for example, safe driving campaigns. In addition, companies will sponsor safety campaigns that may focus on improving behaviors across all work functions and groups.

Continuous improvement within the industrial hygiene discipline often refers to improving the performance of work activities related to the industrial hygiene functional element or a subelement (i.e., exposure assessment process).

1.5 INDUSTRIAL HYGIENE KEY FOCUS AREA AND RISK REDUCTION

There are a number of focus areas within the industrial hygiene program, elements that are key to the industrial hygiene professional, for reducing health risk to the workers and financial risk to the company. These focus areas include:

1 Understanding the application of the art and science of industrial hygiene and how risk may be reduced when evaluating industrial hygiene information.
2 Understanding and targeting risk reduction in the development and maturity of the industrial hygiene professional.
3 Incorporating techniques and methods for reducing risks associated with the management of instruments and calibration processes.
4 Applying risk reduction techniques and methods in the implementation of recognizing, evaluating, and controlling hazards in the workplace.
5 Incorporating risk reduction recognition in the development and implementation of training.
6 Understanding and incorporating risk reduction practices when responding to emergency situations.
7 Recognizing risk reduction opportunities when evaluating the occupational safety and health programs.
8 Recognizing and incorporating risk reduction practices in the development and implementation of a continuous improvement program for the safety and health and industrial hygiene programs.
9 Incorporating lessons learned from enforcement actions that can be used to reduce company and worker health risks.

Consideration of these key aspects by industrial hygiene professionals in the development and implementation of their program can result in reducing both health risks to the worker and financial risks to the company. Finally, adoption of a consensus standard (i.e., ISO 45001) incorporates risk reduction techniques into implementation of the industrial hygiene discipline and goes beyond compliance, but focuses on a risk-based, tailored approach to minimize health risk in the workplace.

QUESTIONS TO PONDER FOR LEARNING

1. Summarize the purpose of the United States Williams–Steiger Act of 1970.
2. What is the relevance of the labor regulations and consensus standards in today's society?

3. Describe the role of Maslow's hierarchy of needs as it pertains to the safety needs of workers.
4. Describe and explain the industrial hygiene program tenets and their importance in the implementation of an effective program.
5. List and discuss the elements of an industrial hygiene program. Are some elements more important than others?
6. What aspects of an industrial hygiene program can risk reduction techniques be applied to?

2 Art and Science of Industrial Hygiene

2.1 INTRODUCTION

The industrial hygiene professional's primary focus is to protect workers from routine and unique hazards in the workplace. Hazards in the workplace generally fall into one of the following categories:

- Physical hazards
- Chemical hazards
- Biological hazards
- Radiological hazards

Today's industrial hygienist is taught various methods and techniques on how to recognize and manage occupational hazards; it is generally a knowledge- and skill-based approach to hazard recognition and control. Each workplace and situation is unique, and the industrial hygienist must develop and adapt scientifically based methods to understand and control hazards. However, when industrial hygienists work in the field, and as they interact with the workers, they must understand how to communicate with workers, which includes gathering, evaluating, and communicating sampling and monitoring results and conclusions. Because industrial hygiene is not an exact science and is influenced by people, professional judgment for assessing risk, and mathematics, it is a profession driven by both art and science. Whether the industrial hygienist is interacting with workers, presenting a summary of operational conditions to management, or evaluating data and determining the level of health risk to the worker, industrial hygiene professionals must learn to balance the application of both art and science into their job. It is the art and science of applying industrial hygiene principles and practices, preventing injuries and illnesses, and the success achieved when working with workers and management that truly makes an industrial hygiene professional successful.

2.2 ART OF INDUSTRIAL HYGIENE

The art of industrial hygiene pertains to the application of professional judgment to everyday work activities and decisions that provide answers to questions such as:

- Do I understand how a defined manufacturing process works?
- Am I familiar with hazards associated with a particular type of process?

DOI: 10.1201/9781032645902-2

- Will the control I selected to mitigate the hazard be accepted by the worker?
- Who should I monitor and why?
- What process should I be monitoring and why?
- What are the monitoring results telling me, and how do they relate to work that is being performed?
- Where should I monitor, is it representative, and will it fulfill the data need?
- How do I effectively communicate to the worker the monitoring results?
- Do the monitoring results thoroughly support conclusions regarding an occupational injury and illness?
- How do I execute work activities when supporting emergency response?

Above are just a few example questions in which industrial hygienists must use professional judgment in their daily decision-making process. The art of industrial hygiene considers:

- Professional judgment of the individual professional
- Conservative decision making, including incorporation of safety factors, as needed, based on potential risk and cognitive reasoning
- Previous work experience
- Feedback and guidance from other professionals
- All aspects of training, such as education, on-the-job training, certification, and company performance expectations

Mathematics supports the definition and data for conclusions, but how the information is evaluated and applied in managing health risk can be largely through professional judgment.

Both a person's experience and personal relationship skills contribute to how industrial hygiene methods and techniques are applied and accepted by others in the workplace. Because the industrial hygienist incorporates art, essentially professional judgment, into execution of the profession, how the information is generated and communicated contributes to the ability of the industrial hygienist to be able to make the worker feel protected, along with effectively protecting the worker. The art and science of industrial hygiene are recognized and incorporated into all tasks performed by the industrial hygienist; however, it is essentially primarily dominant in the following:

- Hazard recognition and control
- Monitoring strategy
- Occupational health program
- People management

2.2.1 Art of Hazard Recognition and Control

Hazard recognition pertains to the cognitive awareness of a person to recognize conditions or situations in the workplace that could cause, or result in, a worker injury or illness. Controlling those recognized hazards, and thus the prevention of an injury or illness to the worker, is essentially the application of hazard controls. Chapters 4 through 6 of this book focus on methods and techniques for recognizing, evaluating, and controlling the exposure of hazards, and risk, to the worker. This section focuses on how an industrial hygienist's education, professional experience, training, and personal relationship skills, the "art" of the discipline, can be applied in the hazard recognition and control process, but also in the management of the overall industrial hygiene program, and when interfacing and building a relationship with the workforce and management.

The industrial hygienist uses a variety of methods and techniques in the recognition and control of hazards in the workplace. Most companies have defined processes by which materials, equipment, or services are produced. The industrial hygienist is required to understand the mechanics of these processes, and in many cases, they have been trained to identify, evaluate risk, and control many of these industrial and traditional workplace hazards. However, there are situations where a hazard may not be easily recognized or controlled, or the worker does not perceive that the hazard or risk has been recognized and/or controlled. In addition, the worker or, as is more often the case, a change in the work activity or task may create a hazard that did not exist when the work process was originally established. For example, a chemical manufacturer may be required to substitute one of its chemical products, used in its manufacturing process, for one that is less toxic. As part of this required change, the company will need to modify its manufacturing process to adapt to a variation in chemical formula and will require an additional piece of equipment. Review of the new manufacturing process, by the industrial hygienist, may identify a new exposure hazard or risk while workers are performing their job assignment, which could result in an increased health risk.

It is the job of the industrial hygienist to review and evaluate a process, and listen to the workers when defining and controlling hazards. The hazard may be relevant to a process that includes multiple pieces of equipment, not just one; but rather collectively, the process itself creates the hazard. In such situations, industrial hygiene professionals may use their experience and personal relationship skills in identifying and controlling the hazard.

For example, a chemical treatment plant may have specific equipment that is designed to remove individual toxic properties of a chemical from the waste stream, and the hazards associated with each treatment unit may be recognized and controlled; however, the synergistic effects of several chemicals together, and their combined toxicological properties and health risks, were not recognized and controlled. Maybe previous experience in a wastewater treatment plant resulted in recognizing the hazards of a treatment process with multiple treatment processes and equipment. Control of hazards to the workers required a routine monitoring frequency and feedback from the workers of whether the hazards have been controlled, based on not only monitoring data but also verbal feedback.

The industrial hygiene professional should also review and consider incidents and lessons that have been learned from a particular workplace, industry, or chemical and apply those lessons in the implementation of a hazard recognition and control program. There are many examples and lessons that can be learned from similar operations and companies that the industrial hygienist may use in improving the recognition, evaluation, and control of workplace hazards.

The Occupational Safety and Health Administration (OSHA) routinely issues a number of publications that highlight management techniques for many different types of hazards. Another excellent resource for understanding causes of chemical accidents is the U.S. Chemical Safety Board, which is empowered by Congress to investigate chemical accidents with the purpose of understanding what caused the accident and how a similar accident can be prevented in the future. The U.S. Chemical Safety Board website (http://www.csb.gov) contains a wealth of information on various industries and how process information can be learned from and applied to other industries.

Along with understanding manufacturing processes, the industrial hygienist must also rely on intuition when recognizing and controlling hazards. Intuition is the inherent ability to recognize and understand something without conscious understanding or reasoning. Intuition is instinctive and is a part of human nature. In many cases, an industrial hygienist may not readily recognize the hazard, but will act to create a protective action because he or she may believe a hazard could be created. It is a proactive response versus a reactive response. In many cases, the use of intuition can prevent an accident or hazard from occurring.

2.2.2 Art of an Occupational Exposure Monitoring Strategy

Once a hazard is recognized, the industrial hygienist must evaluate and determine the method and technique to be used to measure the degree of potential harm the hazard poses to the worker. The determination as to whether a hazard exists may or may not be obvious, but the industrial hygienist will make a decision based on professional judgment as to whether it exists. Depending on the type of hazard and health risk, exposure monitoring is used to collect and analyze data to determine the degree to which the hazard needs to be controlled. Within the United States, most industrial hygienists use a standardized approach to monitoring, such as that defined in *A Strategy for Assessing and Managing Occupational Exposures* (Mulhausen and Damiano 1998). The approach recommended by the American Industrial Hygiene Association (AIHA) is a systematic, defined process whereby characterization of the workplace is conducted, an exposure evaluation is performed, and results of the evaluation are used to determine whether the exposure is acceptable or unacceptable, or more information is needed to make an informed decision on the hazard posed by a particular operational process. At each step of the systematic approach, the industrial hygienist must make professional judgment decisions that will influence implementation of the exposure monitoring strategy.

An exposure monitoring strategy is a defined process for identifying, evaluating, and determining who, what, where, and when to perform monitoring in the workplace. Similarly, in other countries an overall risk assessment approach is used. Chapter 4 presents a more detailed discussion on an exposure monitoring strategy;

however, it should be understood that it is not an exact science; rather, a defined exposure assessment strategy is a combination of art and science.

The identification of similar exposure groups (SEGs) and sampling and monitoring locations uses professional judgment, from evaluation of data and other factors, in the decision-making process. The decision as to what constitutes representative sampling and the periodicity of the sampling and monitoring evolution incorporate professional judgment. Finally, analysis of the data and the determination as to whether an exposure has caused actual harm, or an injury or illness, at times have some level of professional judgment built into the decision-making process.

2.2.3 Art of the Occupational Health Program

One of the responsibilities of an industrial hygienist is to provide input and feedback into an occupational health program. The identification and control of hazards or health risk, the analysis of exposure data, and the determination of ongoing protective actions and monitoring needed to fully understand and protect the worker from physical, chemical, biological, and radiological hazards are essentially the job of the industrial hygienist. How that information is communicated and incorporated into an occupational health program is partly dependent on actions of the industrial hygienist.

An occupational health program is both preventative and diagnostic based on health risks. Typical elements of an occupational health program include:

- Preemployment physical, which is supported by a baseline job task analysis
- Wellness program
- Routine medical monitoring and physicals
- Injury and illness response
- Postemployment physical, which defines and summarizes occupational exposures that have occurred over a worker's employment period

The industrial hygienist, to some degree, contributes to the entire spectrum of the occupational health program. The industrial hygienist works closely with the occupational health provider in communicating and understanding the functions and potential hazards associated with a particular job to further understand health risks. Because industrial hygienists are generally located at the job site, they become the eyes and ears of the medical profession in determining the overall impact that a process or industry can have on a worker.

As the routine medical monitoring program is implemented, the occupational health provider will interface and work with the industrial hygienist to understand the following:

- What hazards and potential exposures could potentially increase a worker's health risk?
- Is there a connection between a physical condition and an exposure?

- Does the worker perceive and communicate that he or she is being protected in the workplace?

There are times when this information is not readily available, or the scientific data may not support a clear link between exposure and a medical condition. The industrial hygienist often uses his or her experience, knowledge, and other professional judgment as a means to assist in a medical prognosis or improve an existing occupational health program. Areas in which the industrial hygienist works with the occupational health provider include:

- Providing process information or conducting work area tours with the occupational health provider to assist him or her in better understanding the work location and potential hazards and controls
- Assisting the occupational health provider in understanding work restrictions or limiting personnel who are sensitized to a particular chemical or exposure scenario
- Communicating environmental and hazardous conditions in case of an emergency
- Understanding health impacts to a particular set of workers and determining whether there is a common exposure mechanism

There are many times in which the industrial hygienist will work with the occupational health provider in identifying and improving potential workplace conditions, but how the two separate professions interact and share information is crucial to working as a team to protect the worker.

2.2.4 Art of People Management

Probably the most important aspect of the industrial hygiene profession is the skill of an industrial hygienist to listen and work with people. The ability of an industrial hygienist to be perceived as actively protecting someone is the foundation by which credibility and trust are built with the workforce. If the purpose of the industrial hygienist is to aid in the protection of people from workplace hazards, and ultimately reduce potential health risks, then he or she must be able to effectively convey what actions are being taken, the methods by which the actions are being implemented, and the results and subsequent health impacts of the actions in a manner that is understood by the workers. In particular, the experience and personal skills of the industrial hygienist significantly influence whether a worker believes and trusts the industrial hygienist and the workplace monitoring data.

There are many companies that have excellent industrial hygiene programs, but whether the workers believe their company truly is committed to safe workplace and that the industrial hygienist is truly interested in their well-being is dependent on the outcome of his or her decision-making demonstrated through daily interactions. Often, the most successful industrial hygiene programs include building a collaborative relationship with the workforce, and the workforce often depends on the industrial hygienist to ensure their safety. Below are the key items for the industrial hygienist to consider in improving their people skills, aka people management.

2.2.4.1 Solicit Feedback from the Workforce

More often than not, the workers have the best understanding of workplace hazards if they have previously performed that work, or have heard from other workers, what are the most hazardous work practices and equipment. The workers themselves possess a vast amount of knowledge, whether it is knowledge- or skill-based; consequently, information received from the workers should always be incorporated into an industrial hygiene program.

2.2.4.2 Team with the Workforce in Preventing Hazards

One person can develop a great plan, but collectively, if more than one person's personality, experience, and knowledge are combined, an even greater plan will be developed. The workers generally have an understanding of how the work is to be conducted and can provide feedback as to how to perform work in a more protective manner. The most productive and safe work evolutions are those that incorporate not just science, engineering, and mechanics, but also feedback and employee engagement.

2.2.4.3 Provide Positive Feedback

Most people come to work wanting to do a good job, and everyone wants to be recognized for their contribution. Respect for the industrial hygienist will improve and grow if the workforce understands that their work and contributions are important to the overall mission of the company and protecting their fellow worker. The industrial hygienist has a great opportunity to promote a healthy working environment because of the interconnected relationship that he or she has with the workforce.

2.2.4.4 Be Respectful and Trusting

It is human nature to not readily trust someone; however, if respect and trust are extended to you, then you are more likely to receive respect and trust. Industrial hygienists must always be aware that they are looked upon as being in a position of authority because they represent the company, and work closely with management in implementation of the safety and health program. They also serve the people and possess the skills and ability to understand whether a worker may have received an occupational injury or illness. It is the responsibility of industrial hygienists to conduct their business in a manner that promotes respect and trust.

2.3 SCIENCE OF INDUSTRIAL HYGIENE

The science of industrial hygiene pertains to the application of scientific practices and principles in execution of an industrial hygiene program. The foundation of an industrial hygiene program is based on mathematics and the interpretation and application of the data. Science and math are critical in determining whether a worker is working in a hazardous environment, and whether the worker has been exposed to a hazardous substance or condition. Math and science are important in determining what concentrations a worker is exposed to, along with formulas to apply when identifying the most appropriate hazard control method. The application of ventilation as a primary hazard control method is a great example of when science is used to determine the required capture velocity for controlling particulates in an airstream.

The exposure assessment process itself is a mathematically driven process that requires the use of math and science in determining the best approach for managing hazardous work environments and identifying occupational injury and illness. The risk assessment, or exposure assessment process, was developed as a scientific approach with defined sequential steps, uses statistics in the sampling and data analysis process to determine a level of protectiveness of the worker to workplace hazards, and is fundamental in the correlation of occupational injuries and illnesses. Data that is gathered through a scientifically based approach assists the occupational health provider in understanding and determining acute and chronic symptoms of an occupational exposure. Although communicating with and managing people is primarily a professional judgment process, assisting people in understanding health risks caused by exposure to a specific concentration level involves the use of math and science.

2.3.1 Science of Hazard Recognition and Control

The science of hazard recognition and control pertains to the application of math and science, by the industrial hygienist, to the identification and management of risks and hazards. The profession of industrial hygiene is dramatically changing in the twenty-first century through the use of the internet and the workforce possessing a greater understanding of hazards in the workplace.

Today's workers, and the general public, are more aware than those 20 or even 10 years ago. If they are concerned about the health effects or risks of a particular chemical, they merely have to search the internet for a plethora of information. Because of the greater availability of information, workers and the general public are developing a lower threshold for accepting risk resulting from occupational injuries and illnesses. Society today believes that the government or company has a greater responsibility for ensuring its workers are not being harmed in the workplace. Consequently, workers are more apt to question whether a particular work evolution or chemical will result in an occupational injury or illness. There exists a lower level of acceptability of risk or exposure, and lawsuits are being seen more frequently, as people are more educated into what hazards may be present in their particular workplace.

The preferred time for implementing an industrial hygiene program starts at the beginning, during the design of a manufacturing process or building, commonly referred to as Safety in Design. If hazard recognition and control practices are initiated during the design phase up to manufacturing, then protection of the worker is inherent to implementation of the manufacturing process.

When implementing hazard recognition and control practices during the design phase, the industrial hygienist works with engineers in understanding what function the process or equipment will perform, what operational evolutions will be performed, and how the maintenance will be conducted. The industrial hygienist, through working with the design team, can then incorporate engineering controls into the system design before the equipment or process is installed, thereby realizing a reduction in the risk posed.

The industrial hygienist is involved in the review of potential manufacturing equipment to recognize what hazards might originate from the equipment itself, but also from any chemicals that may be introduced as part of the manufacturing process. A good example is noise that may be produced when a large motor is procured. By being involved in the selection of the equipment, the industrial hygienist can ensure that additional insulation is added to the equipment to reduce the amount of noise being generated while the machine is operating.

Another example is the procurement of a piece of equipment that requires the addition of ammonia into the process. Product substitution may be an alternative, or the equipment could be located away from workers. Real-time monitoring may be needed and installed, along with a secondary barrier, or water deluge system, should there be an ammonia release. Incorporating industrial hygiene into the design of a system is the preferred method for incorporating safety into the workplace; however, most industrial hygienists do not have the opportunity to be involved in the design of equipment or a process. The industrial hygienist must identify and recognize hazards and risks in a manufacturing facility that uses established equipment. The industrial hygienist did not have the opportunity to provide input into the purchasing of the equipment; rather, the industrial hygienist inherited working conditions, including existing manufacturing and processing equipment, and associated hazards. For most industrial hygienists, the hazard recognition and control process is initiated by understanding how the equipment or process operates and performs, and then applying that knowledge in recognizing potential occupational exposures. Example applications of when the hazard recognition and control process is influenced by science include:

- Particulates are not being adequately removed from a laboratory hood.
- Personnel are injured from a poorly designed control panel on an effluent treatment system.
- Excess smoke and chemicals are being emitted from a process stack to the ground.
- Radiation is being emitted from a calibration detector used in a process system.

There are countless situations that the industrial hygienist will encounter in performing his or her work, but when recognizing and controlling risks and hazards, the industrial hygienist will rely on both math and science in determining the appropriate method for protecting the worker.

2.3.2 Science of an Occupational Exposure Monitoring Strategy

An occupational exposure monitoring strategy is driven by both art and science. The application of math and scientific principles is required when both performing monitoring and analyzing the data. An occupational exposure monitoring strategy includes, at a minimum:

- A statistically representative population for monitoring
- Calibration of monitoring instrumentation
- Collection of monitoring data
- Analysis of the monitoring data into computational data that can be used in evaluating exposure and determining overall health risk to the worker.

Although the evaluation and communication of monitoring results use largely professional judgment, the gathering and conversion of monitoring data into useful exposure data are primarily scientifically based. Statistics are incorporated into an established exposure assessment monitoring strategy as a tool to be used to scientifically prove that the population sampled is representative of the larger worker population.

Computer modeling of contaminants can also be useful in understanding how contaminants behave in a particulate versus vapor form. That information can be useful in the recognition and control of a hazard, as well as in performing exposure assessments. By understanding how the contaminant will behave – both the baseline monitoring and the continuous personal and area monitoring – the industrial hygienist can better depict what conditions the worker may be subjected to and any potential health risk.

Analysis of the data is based on both professional judgment and science. Compiling the data, applying mathematical concepts, and then evaluating the final data results against established risk levels, or regulatory and acceptable limits, are among the primary jobs of the industrial hygienist. The information is then communicated to the occupational health provider, management, and the workforce.

2.3.3 Science of the Occupational Health Program

Science is involved in an occupational health program through the application of data and a clear, scientifically based conclusion that exposure to a particular contaminant or workplace condition is the cause of an injury or illness. Workplace monitoring and the recognition that a hazard is present and occurs over time also involve the use of science. Although most industrial hygienists perform baseline monitoring, depending on the hazard and activity of the workers, the monitoring frequency may be routine and occupational exposure data is collected over time. It is important that the scientific data is accurately documented and communicated.

The data that is transmitted and used as part of a worker's occupational work history is required to be accurate and complete.

There are a number of regulatory requirements associated with industrial hygiene and occupational health programs that support the need for the determination of a workplace injury and illness; this need is driven by the accuracy and completeness of the data. For example, a worker is employed by a foundry that produces steel parts. The occupational history of work exposure to hazardous conditions may demonstrate that the worker was not exposed to sounds in levels above the occupational exposure limits; however, the worker may have significant hearing loss, associated with the work activities, and there is insufficient data collected in the workplace to negate that his hearing loss was not from the workplace.

It is important to the industrial hygienist that workplace evaluations be conducted at a frequency that ensures protection of the worker from hazards, but also that the information is transmitted to the occupational health provider and workers that represents that particular work evolution. Not only is science important in supporting an occupational health program to protect the worker in today's workplace, but also it will be relied on for future litigation and must be sufficient and robust enough to withstand scrutiny over time.

2.3.4 Science of People Management

The science of people management pertains to the ability to demonstrate a clear connection between workers being exposed to a risk or hazard using scientific information, a potential occupational injury or illness, and the understanding of the workers that they were or were not exposed to hazards that may have caused an injury or illness. Management of people is usually not scientifically related. The actions of people, unlike math, are not an exact science and require the industrial hygienist to effectively communicate and understand their perspective. However, there is a science to dealing with people for the industrial hygienist; working conditions and monitoring data associated with work performed must be defensible and credible when communicating and interfacing with the worker. It is the role of the industrial hygienist to ensure that the workers understand that the data collected is protective of their work activities and reliable and defensible.

Since the emergence of computers and the internet, workers have become more educated and have been questioning the protectiveness of workplace air concentrations or exposure to chemicals; they are more sophisticated in their understanding of what constitutes "representative" sampling. The workers are more cognizant of health risks and regulatory limits that are traditionally defined as "acceptable" exposure. This is largely due to the ease with which people can readily look up health hazard information associated with a particular chemical or industry. In today's society, workers can be at work, be told they will be working with a new chemical, look up the chemical real-time on their cellular phone or iPad, or go home and research the health effects and precautions associated with that chemical. They can also easily research whether workplace illnesses are associated with that chemical.

To expand that thought, not only is the information available for how that chemical is managed in the United States, but also everyone has the ability to understand how chemicals and other hazards are managed worldwide. There are several industries in which class action lawsuits have originated and target historical exposures to hazards such as asbestos or beryllium. Because of the large amount of hazard and exposure information (risk) available on the internet, society and the workforce are continuing to increase their level of expectations of protectiveness, and thus are less accepting of an increase in health risks. Today's industrial hygienist must practice scientific techniques for hazard identification and control and data collection practices. Not only is it the right thing to do for the industrial hygienist, but also the depth to which the worker will trust the industrial hygienist is dependent on his or her ability to communicate the math and science associated with the profession.

Questions to Ponder for Learning

1. Discuss the art and science of industrial hygiene.
2. List and provide examples of typical workplace hazard categories.
3. Define *professional judgment* and discuss when and how it is used in the field of industrial hygiene.
4. Explain the process and value of having an exposure monitoring strategy.
5. What are the typical elements of an occupational health program?
6. List and discuss actions that an industrial hygienist can engage in to improve his or her people management skills.

3 Industrial Hygiene Professional

3.1 INTRODUCTION

In the United States, the Occupational Safety and Health Administration (OSHA) mandates that a safe workplace be provided to all workers regardless of industry, product line, or location of the company. Thus, employers have the responsibility to ensure that workers are safe while at work, and that their business practices do not negatively impact the environment. To aid in helping maintain a safe work environment, many companies hire industrial hygiene professionals who have the knowledge to serve as facilitators in achieving compliance with the OSHA mandate, as well as other applicable regulations and consensus standards (i.e., ISO 45001). These professionals are skilled in the anticipation, recognition, evaluation, prevention, and control of environmental factors or stresses that can arise in, or develop as a result of conducting tasks in the workplace, and in understanding the link, if there is one, between exposure to hazards in the workplace and occupational illnesses and injuries (health risk). These stressors and environmental conditions have the potential to cause sickness, impaired health, significant discomfort for workers, and even death.

A professional industrial hygienist is an individual possessing, at a minimum, a bachelor's degree in engineering or a closely related biological or physical science discipline from an accredited college or university, coupled with discipline-specific training. Historically, many of the industrial hygienists were trained on the job and did not necessarily possess higher education. The educational requirement, as well as the knowledge expectations for these professionals, has increased over the years often because of legal litigation and consequently many universities and colleges today have dedicated programs for industrial hygiene. The type of job hazards industrial hygienists are actively engaged in preventing or resolving include:

- Air contaminants: Common contaminants include dust, fumes, fibers, mists, aerosols, and solid particles.
- Chemical hazards: Common hazards are traditionally from solids, liquids, gases, mist, dust, vapors, and fumes.
- Biological hazards: Common hazardous sources include bacteria, fungus, and viruses.
- Physical hazards: Common hazards are from ionizing and nonionizing radiation, vibration, noise, and temperature extremes.
- Ergonomic hazards: Common hazards can result from lifting, pushing, holding, walking, reaching, and gripping.

DOI: 10.1201/9781032645902-3

The field of industrial hygiene is considered both a science and an art because not only is it necessary to use knowledge in the area of science, but also it is oftentimes necessary to use professional judgment coupled with past experiences. The industrial hygiene professional has the skills to enable and facilitate safe work in various environments through the use of science, engineering, and technology.

Industrial hygiene professionals in the twenty-first century is engaged in evaluating a broad spectrum of work environments. In the United States, the OSHA recognizes the field of industrial hygiene and the industrial hygienist as integral parts of facilitating a safe work environment and a healthy work culture. Workplaces are safer as a result of the industrial hygiene professionals applying the principles of industrial hygiene.

The practice of industrial hygiene also known as occupational hygiene has grown and continued to grow around the world. The greatest number of industrial hygienist are working in English-speaking countries such as the United States, Canada, Australia, United Kingdom, Ireland, and South Africa. The profession of industrial hygiene has seen growth in other countries such as Brazil, Sweden, France, Germany, Italy, Spain, Japan, China, and Switzerland. The transition in movement from physician dominated treatment to prevention is one reason fueling the increasing demand for industrial hygienist across the world.[1]

3.2 ROLE OF THE INDUSTRIAL HYGIENE PROFESSIONAL

The fundamental job responsibility of an industrial hygienist is to aid in protecting the worker from workplace hazards and to reduce human health risks. This goal not only includes the workers but also their families and the community. For example, when workers bring home particulate contaminants, such as beryllium, on their clothing, then their family is also exposed to the same health risk as the worker.

Over the years, the industrial hygienist has been viewed as a consultant in the implementation of the safety and health program. He or she has been viewed as a specialist, and only needed when requiring additional skills in the management of the occupational health medical program, or when controlling exposures to toxic air and surface contaminants. Companies that manage their safety and health program in such a fashion are now finding themselves in increased financial and regulatory peril because the industrial hygienist was not recognized as an essential contributor to program development and implementation. In fact, it is the safety and health professionals who are rather limited in their knowledge of hazard recognition and control because their profession primarily focuses on hazards associated with manufacturing – physical hazards, not chemical, biological, or radiological hazards.

The industrial hygiene professional play a critical role in ensuring that federal, state, and local laws and regulations are included in a company's policies and procedures, while management is responsible for ensuring that they are followed. Because of the knowledge and skills that are required, an industrial hygienist can be viewed as a combination scientist and engineer that works to ensure the safety and health of employees while in the workplace. Typical roles of an industrial hygienist typically include:

- Developing procedures and outlining regulations that support workplace safety
- Evaluating the workplace for potential hazards
- Recommending improvement measures to eliminate or reduce workplace hazards
- Evaluating and investigating workplace injuries and illnesses
- Training and educating employees about job-related hazards
- Participating in process and facility design
- Participating in scientific research advancing the field
- Participating in periodic audit of safety and health programs
- Collaborating with occupational health professionals

Worker safety and health is an important task that can be a challenge for corporate leaders. This challenge, if not attended to properly by company leaders with the appropriate level of knowledge of requirements, can lead to catastrophic events that can negatively impact the workers and company. Fortunately, they can hire industrial hygiene professionals to help in understanding regulatory and design requirements to ensure compliance to all of the applicable regulations and consensus standards. This includes understanding safety and health requirements which are enforceable and legally binding. Many corporate representatives have been fined and some jailed as a result of failing to comply with safety and health regulations and to protect their workers. The industrial hygiene professional is key in the plight of assisting companies in the ethical enforcement of these rules and regulations.

3.3 LEADERSHIP AND ORGANIZATIONAL STRUCTURE

Effective leadership is essential to establishing the overall vision of a safe work environment, as well as accurate and responsive communication of that vision. Leadership can be at the senior management level, or as an industrial hygiene manager, lead, or individual contributor. As stated by Tony Robbins,[2]

> leadership is the ability to inspire a team to achieve a certain goal. It's usually discussed in the context of business, but leadership is also how you, as an individual, choose to lead your life. The definition of leadership is to influence, inspire and help others become their best selves, building their skills, and achieving goals along the way. You don't have to be a CEO, a manager or even a team lead to be a leader. Leadership is a set of skills – and a certain psychology – that anyone can master.

Industrial hygienists may be separate from leadership of the organization, but they must work closely together in implementing a safe work environment.

As a leader, it is also necessary to build committed teams with relevant expertise. How an organization is structured reflects the lines of authority, but also how communication flows. The organizational structure must reflect the involvement of experts but also be broadly inclusive of the community in question and establish mechanisms for learning, communication, and open discussion.

An organizational structure can dictate how tasks are allocated and performed, delineate lines, and flow of communication, and define authority and responsibilities to facilitate achievement of the organization's goals. It also determines the manner by which leadership and employees are structurally organized throughout the company and the means by which policies are developed and implemented.

Many companies stress the value of a flat organizational structure. An organization with a flat structure has few, and at times no, levels of supervision between the management and the staff-level employees. In a hierarchical structure, every employee, including the industrial hygienist, with the exception of the president, is a subordinate to someone else within the organization. However, the value gained through a flat structure must be weighed when considering retention of workers, since a flat organizational structure typically does not offer as much advancement opportunities for workers. In some cases, a hierarchical structure may be more beneficial. With either structure, there are advantages and disadvantages that must be considered when implementing the industrial hygiene or safety and health organization. Some of these are listed below.

3.3.1 Flat Organizational Structure

The advantages of a flat organizational structure is that it:

- Increases the employee responsibility level within the organization
- Improves communication between employees and management
- Allows the decision-making process to become easier because there are fewer levels of management
- Reduces the overhead cost for an organization by eliminating the salaries of middle management

The disadvantages of a flat organizational structure is that it:

- Tends to produce a staff of generalists
- Tends to limit the long-term growth of an organization when the effort to maintain the structure exceeds the needs of the organization
- Has less upward mobility for professionals
- Makes it difficult for larger organizations to adapt to the structure without dividing the company into smaller business units

3.3.2 Hierarchical Organizational Structure

The advantages of a hierarchical organizational structure is that it:

- Promotes developing employees as specialists; therefore, they can become experts in their field
- Has more opportunities for workers to advance

- Has opportunities for promotion, motivating workers to perform their jobs well
- Has clear authority and levels of responsibility

The disadvantages of a hierarchical organizational structure is that it:

- Increases the organization's cost due to salaries for multiple layers of management
- Tends to have communication that is disjointed, stovepiped, and less effective
- Promotes the development of employees as specialists
- Increases bureaucracy, which can hinder decisions and organizational and personal growth

When designing the organizational structure, it is absolutely necessary to keep in mind the following:

- Clear lines of responsibility and authority
- Adequate span of control
- Employee development and career advancement
- Clear lines of communication

An example of an organizational structure designed with providing career growth for the industrial hygiene professional is shown in Figure 3.1. The structure in Figure 3.1 represents an example for a large company with a large staff. The structure can

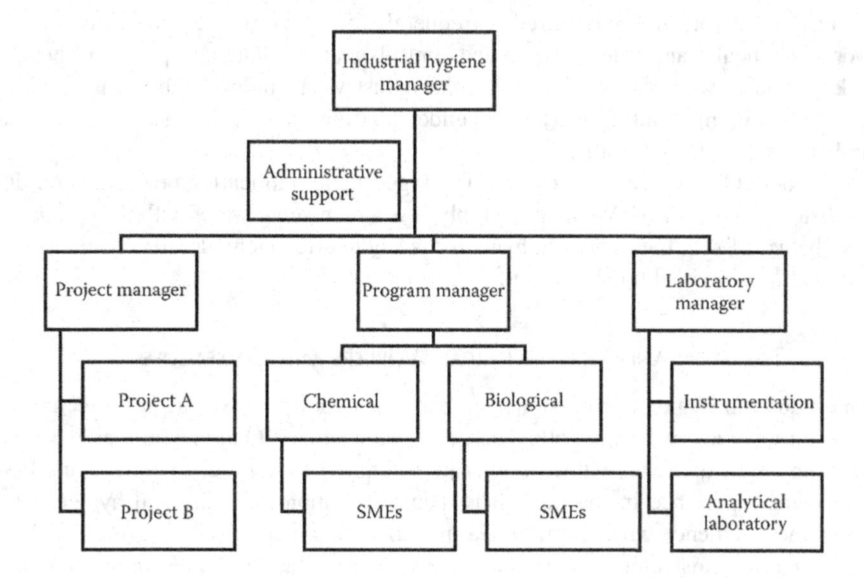

FIGURE 3.1 Industrial hygiene organizational structure for career growth.

be modified and adapted for a small organization as well. The point is that when designing the organizational structure, retention and advancement opportunities must be a factor, along with all other factors that render an organization successful.

The structure of an organization, along with the style of the leadership team, can serve as a positive facilitator of employee engagement and retention. In order for industrial hygienists to be successful in their roles, they must be engaged in business processes having knowledge of where employees may be potentially exposed to hazards or become injured.

Time and time again, we hear from others that "knowledge is power." This certainly is true for industrial hygienists since knowledge in business processes, hazard recognition, and mitigation is key in ensuring that work is done safely. This knowledge makes the industrial hygienist a valuable resource that must be retained.

3.4 COMMUNICATION FOR RESULTS

Effective communications can often make or break the relationship an industrial hygienist may have with management, workers, and the public. As defined by the Oxford Dictionary, communication is the imparting or exchanging of information or news. For the industrial hygienist, communication is the primary means by which interactions with workers and management occur.

The industrial hygienist may be reviewing a work package to identify hazards, and they are expected to effectively communicate the information regarding potential hazards and control measures in a timely manner. The industrial hygienist typically communicates results of monitoring for contaminants, and it is the industrial hygienist who must work with and communicate with the occupational medical professional and the worker in managing the prevention of physical harm. As an industrial hygienist you will be required to frequently meet with workers to communicate workplace health and safety profile and controls and facilitate acceptance of health risk associated with working in a particular industry. The industrial hygienist is vital in facilitating and leading workers in understanding and accepting the health risk and how to protect against it.

As quoted by Stephen R. Covey,[3] "The biggest communication problem is we do not listen to understand. We listen to reply." As a representative of both the company and the workforce, the industrial hygienist is key to effectively identifying and communicating both health hazards and mitigation risks.

3.4.1 Technical and Non-Technical Workers and Colleagues

The industrial hygiene professional interacts on a daily basis with both technical and non-technical workers, colleagues, and management. One person may have 30 years in field experience while another person may be fresh out of college and has minimum experience in the field. It is incumbent upon the industrial hygienist to know their audience when communicating hazard and health risk information. The industrial hygienist should communicate in a manner that facilitates an understanding of the information being presented. One method to employ when communicating health risk information is to compare the health risk for the proposed work activity against health risks associated with work activities away from work.

For example, if workers are assigned to use chemicals when performing maintenance on industrial equipment then the industrial hygienist could compare the health risk associated with performing the work (including the use of personal protective equipment) against performing maintenance of equipment used at home. Often people do not realize that the work they do at home can be more hazardous than work performed in the workplace because of the requirements at work to assess and mitigate health risks. In communicating the health risks to workers it challenges them to consider improving their own safety at home as well as in the workplace. The use of pictures or communicating historical events is also helpful when communicating to both technical and non-technical workers. The level of detail needed to effectively communicate key points should also be considered depending upon whether communicating to the workforce versus management.

3.4.2 Risk/Exposure Assessment Data and Results

The risk assessment and/or exposure assessment process is the primary method used by the industrial hygienist to understand health risk and identify associated mitigation strategies. Consequently, how effectively the information is communicated and managed can have a significant impact on how well the worker or management accepts the information. It is important the industrial hygienist recognize that people exhibit different levels of risk acceptance, and how the industrial hygienist communicates the risk will drive risk acceptance. In addition, the industrial hygienist is also considered a representative of the company and how information is communicated related to exposure risk can influence the degree to which the worker trusts the company and their management. To enhance how risk information is communicated, companies will often develop special training for the industrial hygienist on how to interact and communicate health risk information.

3.4.3 Relationship with the Workforce

As a representative of the company, the industrial hygienist is instrumental in managing health risks posed by operations and acceptance of the risks and mitigation strategies by the worker. The industrial hygiene professional can apply the following known attributes to build a positive relationship with the workforce (Alston, 2014)[4]:

- Openness and honesty: This attribute is frequently cited as the most important attribute because together they form the foundation for trust. If the industrial hygienist is not perceived as being open and honest, then the relationship will not progress in a health manner.
- Competence: The ability to demonstrate competence is dependent upon the industrial hygienists' ability to make good sound decisions on risk mitigation.

- Concern for employees: Industrial hygienists, who demonstrate concern for employees, are well on their way to developing a strong relationship with the workforce. Most people can recognize when a person is genuine and truly cares about their welfare vs. those who are focused on getting the work done to ensure their own success and do not truly care, or ingenuine, about others.
- Identification: Trust is facilitated when individuals share an association with or assumption of the qualities, characteristics, or views of another person or group. The industrial hygienist is often part of the work planning group, and thus can facilitate recognition by others that they are part of the team.
- Reliability: Reliability means not only showing up for work, but also someone who can be counted on for specific roles and responsibilities. The worker relies upon the industrial hygienist to be honest and consistent with communication of health risk information and mitigation strategies.

How well the industrial hygienist exhibits these attributes can influence whether the relationship with the workforce will be positive or contribute to an unhealthy safety culture.

3.4.4 Engagement on Work Planning Teams

The industrial hygienist is typically considered an integral part of the work planning team. The industrial hygienist is frequently involved in design activities to incorporate engineering controls into system design. Often engineers look at the mechanics of the system, but not necessarily the ability to effectively mitigate risk to humans when designing. Included with identification of engineering controls is the understanding of the health risk that can be posed by an operation, requirements that need to be incorporated into the system design, and level of mitigation desired. Once the system is operational, the industrial hygienist becomes a member of the operations team, but also serves as a company representative for day-to-day operations providing safety and health support. The industrial hygienist may have a role in the planning of work, or they may perform sampling, data evaluation, and communicate sample results (and health risk information) to the worker. The role of the industrial hygienist is ever evolving and should be recognized as a required member of the work planning team.

3.5 CORPORATE PROGRAMMATIC SUPPORT ROLE

Nearly all companies have developed a corporate position on how the safety and health of the workforce are valued and supported. It is commonly recognized that safety and health programs and policies will not be successfully implemented without senior management and corporate support. In order for a safety and health philosophy and program to progress and be viewed as an integral part of the corporate program, the management team must lead and demonstrate their support by:

- Providing the financial support needed for safe execution of work
- Providing necessary staff to support the program
- Training of subject matter experts (SMEs) and workers
- Providing technology and needed resources
- Demonstrating safety leadership in the way they conduct business
- Facilitating a learning culture so that workers and supervisors are able to learn from mistakes.

3.6 INDUSTRIAL HYGIENIST AS AN EXPERT WITNESS

The role of an industrial hygienist is more expansive and important than one may think. Not only are these professionals key in developing programs that are designed to minimize the potential for employee injury, they also may be the individuals serving as an expert witness in a court of law. To function well as an expert witness, the professional must be of sound ethical character and viewed as such, and they must possess technical knowledge and experience in the entire field of industrial hygiene. To be effective as an expert witness, the professional must have credibility in decision making and the ability to communicate technical information to technical and nontechnical individuals. Generally, the role of an expert witness is not discussed during the educational process for the industrial hygiene professionals. This is considered a gap in the training for these professionals and must be addressed. Educational institutions are advised to consider including ethics training, along with boardsmanship and the protocol for serving as an expert witness.

3.7 CONTINUING EDUCATION AND PROFESSIONAL DEVELOPMENT

There are several colleges and universities that have degree programs to develop industrial hygiene professionals, although in recent years many of the programs have changed or are no longer in existence. Furthermore, some of the curricula offered by these universities are not very comprehensive and may not meet the needs of employers as they seek to fill their professional positions.

The American Industrial Hygiene Association (AIHA) publishes on its website (https://www.aiha.org) a list of colleges and universities that offer degrees that are accredited by Accreditation Board for Engineering and Technology (ABET) in the discipline of industrial hygiene. ABET is a nonprofit organization that accredits colleges and university programs in the area of science, engineering, and technology. This accreditation provides some assurance to employers that a college or university program meets the quality standards that were established by the profession, and the program has the curriculum to prepare the student to be successful in the discipline. Many employers are seeking to hire candidates from ABET-accredited universities or colleges. The list provided is a great start in identifying programs designed to develop industrial hygienists.

3.7.1 College and University Curricula

Colleges and universities play a critical role in educating students that will someday take on roles as safety and health professionals. It is even more challenging for these institutions to develop and support curricula that are suitable in educating future industrial hygienists. Industrial hygiene is a niche field, and many institutions over the years have changed their programs to suit a wider range of professionals in the environment safety and health (ES&H) field. This change was necessary for some institutions because the industrial hygiene field of study could not sustain the necessary student enrollment for these institutions to justify maintaining the program.

A good college curriculum has a good balance of science, math, technology, and on-the-job training (internships). Some of the critical courses that should be a part of a curriculum seeking to educate and prepare students to be successful industrial hygienists include specialized courses, such as the ones listed in Tables 3.1–3.5. To fully round out a successful course of study, the addition of soft skill courses will further help develop and propel the student into the workplace. Some of these classes should include, at a minimum, the courses listed in Table 3.5. Elements of a good progressive college program are shown in Figure 3.2. Therefore, when selecting a college or university that is expected to provide the knowledge needed, one must

TABLE 3.1
Key Industrial Hygiene Courses: Technical Knowledge

Course	Knowledge Objective
Industrial hygiene	Build theoretical and practical knowledge in the area of industrial hygiene
Toxicology	Aid in the understanding of the adverse effects of chemicals on humans and the impact of chemical, biological, and physical agents on the body
Epidemiology	General knowledge of this field can be used to help plan and develop strategies to prevent workplace illnesses and manage workers who may have been exposed or developed disease
Research methods	Expose students to the various research methods that can be used when evaluating issues needing novel solutions
Occupational safety and health	Provide a general yet comprehensive overview of the field of occupational health
Safety and health management	Introduce students to leadership concepts
Industrial hygiene internship	Too often students will spend their time studying hard and focusing on graduating, and not taking the time to gain experience in the field through internships; internships provide students with a look into what a typical day working in the field looks like; they also provide a means to gain experience that can be used as work experience on a resume when seeking employment upon completing their course work
Ventilation systems and engineering controls	These systems are key in controlling exposures to fumes and vapors that can be present as a result of operating parameters

TABLE 3.2
Key Industrial Hygiene Courses: Mathematic Reasoning

Course	Knowledge Objective
Industrial hygiene calculations	Provide the skills and knowledge needed to complete complex calculations to quantify exposures
Statistics for engineers	Provide the skills and knowledge needed to calculate and interpret data to demonstrate exposure potential and potential risks

TABLE 3.3
Key Industrial Hygiene Courses: Monitoring and Analytical Method

Course	Knowledge Objective
Sampling and analysis	Personal monitoring data collection and interpretation to quantify exposures or the lack thereof
Industrial hygiene laboratory	Exposure to the various analytical methods and provide skills needed to select, direct, and interpret laboratory analysis data
Industrial hygiene equipment and instrumentation	Provide knowledge needed to select, calibrate, and utilize sampling equipment to collect personal air monitoring data

TABLE 3.4
Key Industrial Hygiene Courses: Hazard Recognition and Analysis

Course	Knowledge Objective
Environmental health risk assessment	Exposure and knowledge of the risk assessment and decision-making process
Industrial hygiene control methods	Exposure to methods used to control hazards and reduce potential exposures

TABLE 3.5
Key Industrial Hygiene Courses: Soft Skills

Course	Knowledge Objective
Communication • Verbal • Nonverbal • Technical writing	Industrial hygiene professionals are successful when they possess good communication skills that will provide them the ability to communicate challenging and sometimes unpleasant information to management and workers
Presentation	Provide insight into how to deliver an effective presentation that is clear and can be understood by the audience receiving the information
Dealing with conflict	Insight into how to handle effective communication with employees and leadership

College curriculum	Corporate partnership	Work experience
• Math • Science • Soft skills • Technical specialized courses	• Curriculum development • Internships • Special project opportunities • Mentorinship	• Internships • Shadowing • Special research projects with external institutions

FIGURE 3.2 Elements of a good progressive college program.

ensure that the educational institution has a curriculum that will truly prepare the professional for success.

There are many programs advertised that are not considered comprehensive and will not provide the breadth and depth of knowledge needed and, subsequently, may not be considered credible by some or many employers. In such cases, the graduates will become frustrated because they are unable to land a job in the profession which they believed they had prepared for during their university studies.

3.7.2 Retention of the Industrial Hygiene Professional

Retention of workers is a continuous challenge for many companies as they continue to manufacture products and discover novel technologies and ways of conducting business. Worker retention is also more challenging during times when the economy is flourishing and providing options for workers. There are several actions that can be helpful in retaining industrial hygiene professionals. Some of these are listed below:

- Provide financial compensation
- Provide advancement opportunities
- Support training and certification
- Provide opportunities to rotate assignment at least every three to five years
- Implement practices that will provide workers a sense of ownership and belonging
- Recognize workers for the work they do and establish a positive work environment
- Support career aspirations
- Treat everyone fairly
- Ask for and listen to their opinions

The three top predictors of employee retention for an industrial hygiene professional are shown in Figure 3.3. These elements are critical and deserve close monitoring by

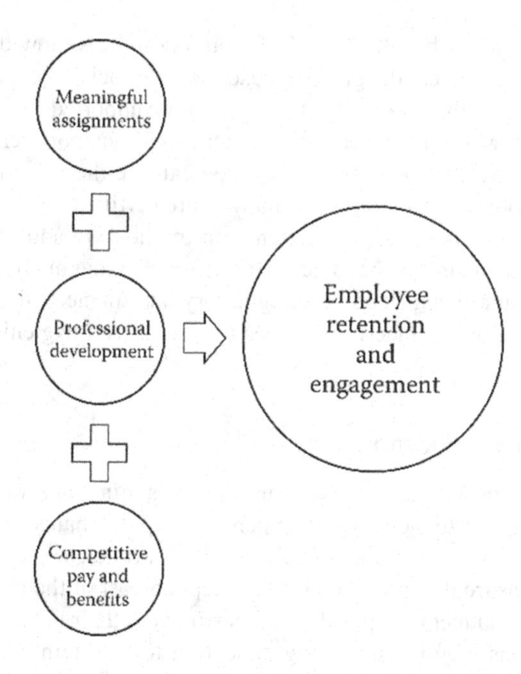

FIGURE 3.3 Industrial hygiene retention and engagement predictor.

management. If these elements are not internalized and facilitated through programs and policies and internalized by these professionals, one can certainly expect turnover to occur at a potentially modest to high rate.

3.7.3 Industrial Hygiene Certification

Certification is a natural progression for many professionals who have a desire to achieve an increased level of pay and/or career advancement. Not all professionals will seek to become certified; however, for those who become certified, they generally have additional opportunities available to them. Certification signals to an employer that an individual meets the highest level of academic and professional standards for their profession. For many professionals, certification is a step that is taken within three to five years of completing their college degree and having spent time working in their career field. Employees who receive support, and even sponsorship, from their leadership team to obtain and maintain their certifications may elect to remain with the company because not all leadership teams and companies value certification for professionals.

Many employers are seeking professionals who will seek certification as soon as they have obtained the necessary years of experience needed to sit for the exam. Many companies value certification and will tend to invest in assisting the professional in achieving it. Some employers will send the industrial hygiene professional to training, allow time to study during the workday, or even pay for exam fees.

Once certification has been achieved, the big question is, how do you maintain certification? This can be challenging because of the associated expense that can be imposed on professionals, especially if they are not supported by the organization that employs them. Many managers or companies will support certification maintenance and the associated costs as long as they can see the value to the company. Therefore, it is important that professionals who are certified display a high level of professionalism and demonstrate competency in the field. In addition, those industrial hygiene professionals who become certified are also sought after when the resolution of issues is significant (increased regulatory and financial liability) and when companies and institutions face legal issues (the industrial hygienist can act as an expert witness).

3.7.4 Continuing Education

Technology and industry practices are constantly changing, and workers need continuing education or professional development to adapt to changes in the regulatory requirements and work environment and retain their proficiency. Continuing education is a must to ensure that professionals are kept abreast of the changes in regulatory requirements and technologies that are pertinent to their field.

Many professions require continuing education to maintain current knowledge and obtain new knowledge required to facilitate compliance with laws and regulations or to maintain licenses and certification. Continuing education is a practical way for professionals to keep abreast of the changes in their fields and increase their education and value. Many employers have a program in place to assist workers in continuing their education and keeping current with the changes in their respective field or profession.

3.7.5 Job Rotation

Job rotation is an important philosophy to consider when evaluating the job satisfaction and retention of industrial hygiene professionals. Some professionals tend to select a career and stick with it, and at times the same position, until they decide to retire. Most people, however, will at some point seek variety in the work they perform daily – not to say that the industrial hygiene professional daily roles are the same from day to day. The typical day for these professionals can range from one end of the spectrum to the other. However, an industrial hygiene professional who is responsible for one project or process will likely become bored and eventually seek the opportunity to gain knowledge and experience in other projects, processes, or aspects of the field. Thus, developing a retention strategy that includes job rotation is a must if there is an expectation that the professional will chose to have some level of longevity with a company. An example of how job rotation can be structured is found in Figure 3.4.

Figure 3.4 represents an example of the type of creativity that management must engage in to retain and develop their industrial hygiene professionals, recognizing that some professionals may have the ability to advance sooner than others. In

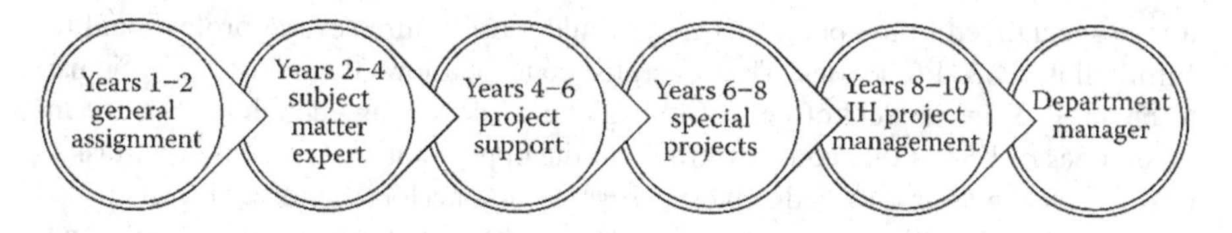

FIGURE 3.4 An example of industrial hygiene (IH) professional job rotation.

addition, some professionals may have a passion for being engaged in specific types of tasks only and may not be interested for example in a leadership role.

3.7.6 Industrial Hygienist as a Generalist

There are times when some industrial hygiene professionals can become bored and lose passion for the discipline because the work performed is so specialized. Because of this specialization, these professionals find themselves focusing on the same types of concepts and issues day after day. Therefore, it is incumbent upon managers to provide opportunities for these processionals to gain additional knowledge and skills that can be used to continue to challenge them, as well as serve the workers, organization, and company. The term *generalist* is used to define an industrial hygienist that has skills in other areas of ES&H and can perform some functions in other ES&H disciplines. One good concept to consider is to develop and implement a training program that will equip these professionals to effectively function in others aspects of ES&H. An example of what that growth opportunity can look like is shown in Figure 3.5.

The ES&H generalist concept has been gaining traction in recent years as a means for companies to hire an individual capable of performing a wide array of ES&H functions. Not only is this concept providing a way to diversify professional skills, but also it can save a company from investing additional funds for workers that

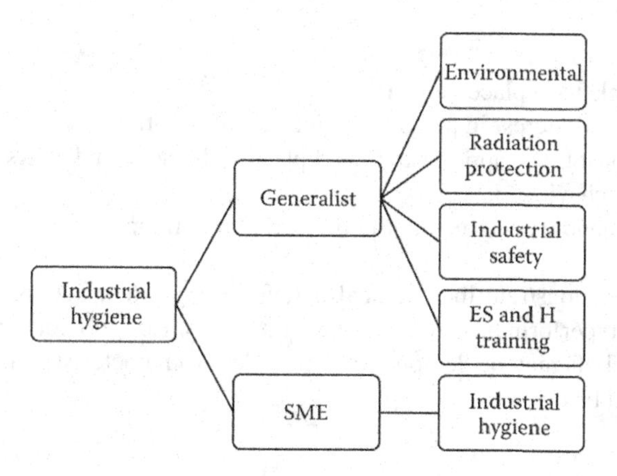

FIGURE 3.5 ES&H generalist concept.

are so specialized to the point that they would have to hire several professional to fulfill all its ES&H functions. The generalist concept allows the industrial hygienist to be an SME in the field of industrial hygiene while having enough knowledge in other areas of ES&H that could afford them the opportunity to perform other tasks, thus expanding their knowledge base. Learning and performing other ES&H tasks can provide the challenge needed to keep the professional interested in continuing to work for a company.

3.8 LEGAL AND ETHICAL ASPECTS OF INDUSTRIAL HYGIENE

The ethical business dealings of professionals and companies have and continue to receive focus across the globe. This is partly because of all the reports and discussions by many on the unethical decisions and actions of company leaders and professionals. With most professions, there are legal considerations that must be taken into account while making decisions and carrying out their daily functions. The industrial hygiene profession is no different in that there are a plethora of decisions that touch on the legality and ethical handling of issues in support of personnel health and safety. These considerations, both legal and ethical, must be applied daily to the decision-making process. Because the work of an industrial hygienists is mostly driven by regulations, such as by HSE, OSHA, and the EPA, they must be knowledgeable in many regulatory requirements and vigilant in their interpretations of these regulations. If regulations are interpreted incorrectly, there is a risk of noncompliance, injury to personnel, or lost reputation of the industrial hygiene professional. The ethical conduct of these professionals can make the difference between whether the information and advice they provide is believed or dismissed.

The American Board of Industrial Hygiene (ABIH) sets the standard and expectations of the ethical performance of industrial hygiene professionals, especially those professionals who are certified by the board. In fact, the board began requiring ethics training in 2010 for all applications and certified industrial hygienists who are seeking recertification. Comprehensive ethics training may include training in the areas of:

- Acceptable workplace behavior
- Conflicts of interest in performing job responsibilities
- What constitutes appropriate workplace behaviors and ways to handle adverse behaviors
- Ways to handle disagreement in professional opinions

Professionals demonstrate their ethical beliefs through their actions and reactions during their job performance and dealing with others. Key components and ways in which these beliefs may be demonstrated include the characteristics listed in Figure 3.6 and defined below.

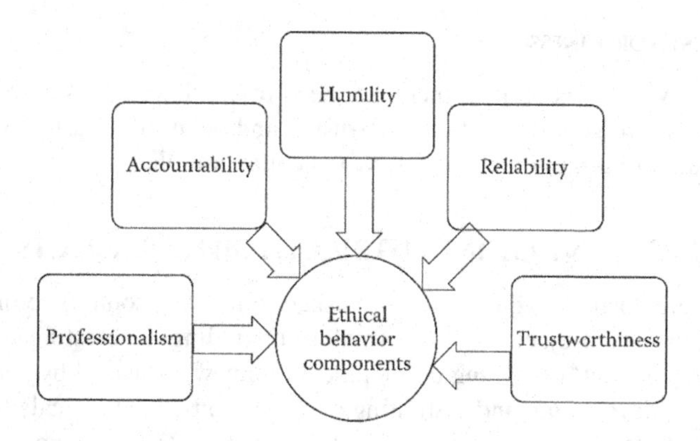

FIGURE 3.6 Ethical considerations for an industrial hygienist.

3.8.1 Professionalism

Demonstrating professionalism involves and encompasses every action of individuals, from the way they dress, to the way they present themselves to others, to the way they treat others. Professionalism is at the core of a strong work ethic. Professionalism can be demonstrated through behaviors such as projecting a positive attitude, treating everyone with respect, engaging in ethical transactions, and being honest in all actions and transactions.

3.8.2 Accountability

Accountability is demonstrated through taking personal responsibility for your actions and outcomes while avoiding making excuses when things do not go as planned. Mistakes are used as teachable moments to prevent reoccurrence. Leaders also expect employees to meet the same high standards and support those who accept responsibility and do not blame others when mistakes are made.

3.8.3 Humility

Humility acknowledges everyone's contributions and freely shares credit for accomplishments. Individuals possessing it frequently show gratitude to colleagues who work hard and appreciation for their support and contributions. Every person is viewed as a valuable contributor to mission success.

3.8.4 Reliability

Reliability is demonstrated through keeping promises and being consistent in responses to and treatment of others. It means having the respect and trust of others. A reliable individual is typically trusted and is respected by peers and management because they are viewed as a dependable team member.

3.8.5 Trustworthiness

In order to be viewed as ethical, there is a need to be viewed as trustworthy. Your word or actions are genuine and can be trusted and believed. Trustworthiness is a key component of building team cohesion and collaboration.

3.9 MANAGEMENT OF INDUSTRIAL HYGIENE PROJECTS

Over time the industrial hygienist may be asked to manage some type of industrial hygiene project. The project may be as small as upgrading the existing contaminant monitoring equipment, managing development of a new industrial hygiene manual, or researching, developing, and instituting a process to track chemicals from cradle to grave. While managing projects there will be a need to use some project management concepts and tools. For example, a project description should develop and describe the problem (may deploy concepts such as brainstorming, process mapping techniques, work observations, employee interviews). A scheduling tool may be necessary depending on the size of the project and the involvement of various resources.

As depicted in Figure 3.7, skills that are important to the industrial hygienist in managing projects and programs include (Alston and Millikin, 2016)[5]:

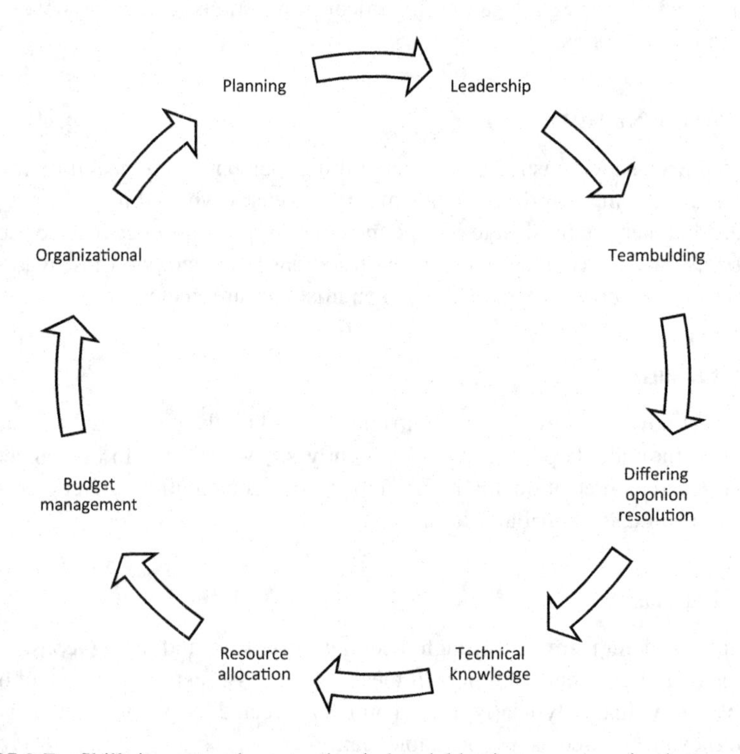

FIGURE 3.7 Skills important in managing industrial hygiene programs/projects.

- Leadership skills
- Team-building skills
- Differing opinion resolution skills
- Technical knowledge
- Resource allocation skills
- Budget management skills
- Organizational skills
- Planning skills

Industrial hygienists often serve in multi-faceted roles and as such, courses associated with project management are recommended to develop a well-rounded industrial hygiene skill set.

3.10 EMERGENCY RESPONSE AND MONITORING

As the acceptance of health risk in the workplace has decreased, the role of the industrial hygienist has increased in their level of involvement in responding to emergencies and communicating the level of risk posed by an emergency. The health risk may be to workers at a localized location, or the health risk may be to a community from a chemical accident. The industrial hygienist is responsible for developing and executing a strategy for monitoring workers and the public but is also responsible for collecting and evaluating data to understand health risk. Training in roles and responsibilities of the industrial hygienist, during an emergency, is key to effectively respond to further reduce health risks from an event.

QUESTIONS TO PONDER FOR LEARNING

Choosing a career is a big decision for high school graduates. Once individuals decide the career that they would like to pursue, they must begin the process of preparing themselves to take on the role and responsibilities of the position, and then secure a job in their profession of choice. The same is true when selecting the discipline of industrial hygiene. In preparing for and excelling in a career as an industrial hygienist, the following questions should be considered and explored, at a minimum:

1. When given the choice to take classes to complete your industrial hygiene program of study early or taking on a summer internship, which do you believe would be the best choice? Why?
2. What can employers do to provide the industrial hygienist with an opportunity to function in other roles, as opposed to having the professional remain in the same position and with the same project for an extended period of time, which may include his or her entire career?
3. Why should industrial hygienists seek to diversify their skills when the opportunity arises?
4. Outline what is believed to be a comprehensive preparation strategy that should render an individual a highly skilled industrial hygienist.

5. What role can a company and the leadership team play in ensuring that the professional has access to continuing training and development opportunities once hired?
6. Describe the role of an industrial hygienist in the work environment.
7. Describe the attributes of a good college curriculum.
8. List and describe key industrial hygiene courses that are designed to prepare the professional for career success.
9. List and discuss the top three predictors of employee retention and engagement.
10. Discuss the importance and role of certification and continuing education in ensuring that the industrial hygiene professional is current in the profession.
11. Summarize the legal aspects of industrial hygiene.
12. List and discuss the components of ethical behavior that should be exhibited by the industrial hygiene professional.

REFERENCES

1. Swaminathan GS, 2014. September–December. Industrial Hygiene: A Global Perspective. *Indian Journal of Occupational and Environmental Medicine*, 18 (3): 103–4. doi:10.4103/0019-5278.146904. PMID: 25598612; PMCID: PMC4292192
2. Robbins T, 2023. *Tony Robbins Website.* https://www.tonyrobbins.com/what-is-leadership/
3. Covey SR, 1988. *The 7 Habits of Highly Effective People: Powerful Lessons in Personal Change.* Simon and Shuster.
4. Alston F. 2014. *Culture and Trust in Technology-Driven Organizations.* CRC Press.
5. Alston F, Millikin EJ, 2016. *Guide to Environment Safety & Health Management: Developing, Implementing, & Maintaining a Continuous Improvement Program.* CRC Press.

4 Strategies for Exposure Monitoring and Instrumentation

4.1 INTRODUCTION

Every day workers are exposed to hazards, and subsequent risk, that needs to be quantified and understood. One of the primary responsibilities of the industrial hygienist is to identify, evaluate, and determine overall health risk to the worker from exposures to physical, chemical, biological, and radiological contaminants. The need to determine worker exposure applies across all work activities, whether a secretary is working on her computer in the office, or a mechanic is performing preventive maintenance on a piece of equipment.

The potential exposure to hazardous contaminants, and subsequent health risk, is required to be evaluated and documented. Listed below are just a few reasons why exposure monitoring should be performed:

- Occupational safety and health regulations require exposure monitoring to be performed, and compliance to applicable standards must be demonstrated.
- Engineering controls should be verified.
- Workers' compensation costs can be reduced.
- Workers have the right to understand what health risks they are exposed to when performing their job functions.
- Companies have the responsibility to understand health hazards and risks created in the manufacturing of their product.
- The selection of personal protective equipment must be determined and technically justified.
- Workers can be assured that the workplace is safe.
- The containment or removal of contamination can be demonstrated.

The exposure assessment process is the primary method used by the industrial hygienist to evaluate, quantify, and protect workers from environmental and workplace contaminants that can cause acute and chronic health problems. The American Industrial Hygiene Association (AIHA) has established a well-defined exposure assessment strategy that is used today by most industrial hygienists. Figure 4.1 is a simplified diagram of an exposure assessment model.

DOI: 10.1201/9781032645902-4

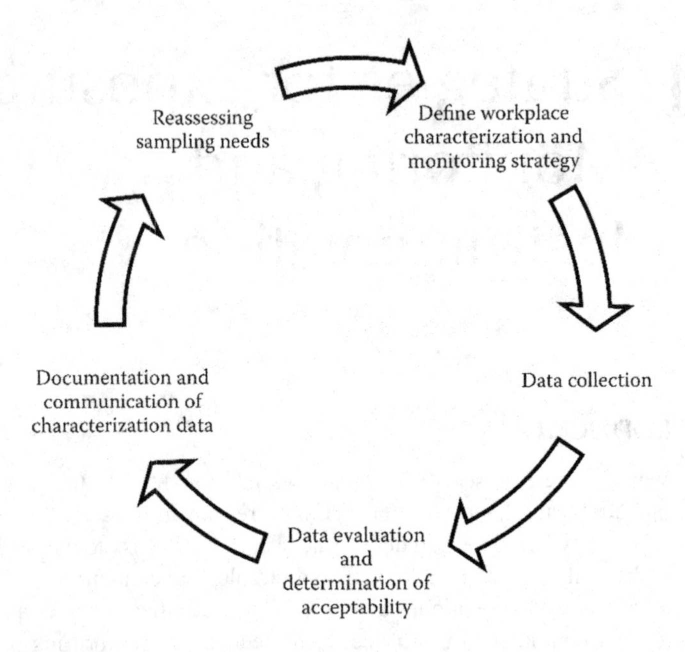

FIGURE 4.1 Example of an exposure assessment model.

Key aspects of the exposure assessment model include:

- Defining workplace characterization and monitoring strategy
- Collecting data
- Data evaluation and determination of acceptability
- Documentation and communication of characterization and exposure data
- Reassessment of sampling needs as required

The exposure assessment (monitoring) process is both qualitative and quantitative. Industrial hygienists must rely on both their experience in making decisions (professional judgement) regarding hazards posed by contaminants, and data analysis and statistics in determining and assigning worker exposure. The industrial hygienist must also recognize that the exposure assessment process is iterative; workplace conditions and hazards can frequently change, along with the manner by which work may be performed.

Fundamental to appropriate implementation of the exposure assessment strategy by the industrial hygienist is the completeness of characterization information to comprehensively address health risks posed by contaminants, ensure the correct sampling methods and instrumentation are appropriately used, and that the entire spectrum of contaminants of concern and hazards has been evaluated and mitigated. The ability to effectively communicate this information is extremely critical for the industrial hygienist to build trust and credibility among the workforce.

4.2 REGULATORY ASPECTS OF INDUSTRIAL HYGIENE MONITORING

Many regulations have been written, or are in the process of being promulgated, which dictate the type of monitoring. As the public awareness of, and education in, the adverse health effective of exposure to contaminants increases, so too must the knowledge and communication skills of the industrial hygienist to be proactive in being compliant with all applicable regulations. Many of these regulations require employers to assess activities of the work environment and provide documentation of monitoring results. The exposure assessment process is required for those activities that are likely to cause exposure to employees or visitors to hazardous substances, such as solids, liquids, vapors, gases, mists, and radiological and biological agents. Some regulations have specific requirements for the type of exposure assessment and time frame when monitoring is expected to be performed in order to demonstrate compliance to a regulatory limit.

When evaluating a work environment, industrial hygienists should ensure that they are aware of all regulatory drivers associated with the contaminants of concern and associated industry and health risks. For example, the industrial hygienist who works in the environmental remediation field must be cognizant of not only safety and health regulations that pertain to performing specific work scopes, such as asbestos, lead, and mercury monitoring, but also safety and health regulations pertaining to a hazardous waste operations and emergency response (29 CFR 1910.120). Noncompliance to a regulatory requirement can be costly to both employers and employees, and the associated costs can go far beyond the obvious financial aspects. Additional impacts and costs can include:

- Increased worker compensation insurance rates and potential overexposures to workers (resulting in an increased health risk).
- Lack of trust by employees.
- Lack of trust from the public and stakeholders.
- Increased regulatory inspections and oversight.

Many chemicals have defined occupational exposure limits (OELs) (e.g., Occupational Safety and Health Administration Permissible Exposure Limits). An OEL is the upper limit that workers can be occupationally exposed to in the workplace. The OEL is typically time and concentration dependent.

Countries all over the world have collectively worked together in developing a process for establishment of international OELs. For example, the Organisation for Economic Co-Operation and Development (OECD) published a report in June 2022,[1] which summarized results of a survey of OECD stakeholders on OEL derivation activities, with the goal of highlighting similarities and differences. The document presented roles, responsibilities, and scope of the responding organizations, methods of OEL development; successes and challenges of OEL development, and interest in and potential areas of focus for international harmonization of OELs. Another example of collectively establishing international OELs is an article from the *Journal of Occupational and Environmental Hygiene*,[2] "The Global Landscape of Occupational Exposure Limits – Implementation of Harmonization Principles to Guide Limit Selection," which explores the underlying reasons for variability in OELs and recommends the harmonization of risk-based methods used by OEL deriving organizations. A framework is also proposed for the identification and systematic evaluation of OEL resources, which industrial hygienists can use to support risk characterization and risk management decisions in situations where multiple potentially relevant OELs exist.

For those chemicals that do not have defined OELs, occupational safety and health professionals in the United States can use the National Institute of Occupational Safety and Health (NIOSH) occupational exposure banding process. It allows users to quickly and accurately assign chemicals into specific categories (bands); enabling industrial hygienists to make timely hazard control (risk-based) decisions based on the best available scientific information. Internationally, other countries use a similar approach to NIOSH's control banding process in determining health risk.

4.3 QUANTITATIVE AND QUALITATIVE EXPOSURE AND RISK ASSESSMENT

There are two accepted methods to conduct workplace characterization. The two methods refer to either a quantitative and qualitative approach to collecting and evaluating data. Both methods are needed to ensure accuracy of the data, proper quantification of risk, and development of risk mitigation measures. The preference should be to conduct exposure or risk assessments using a quantitative approach or a combination of the two; however, whatever workplace characterization approach is chosen, the logic behind the selection must be documented in case there are questions from the workforce on their exposure profile and legal questions in the future regarding a potential overexposure to a contaminant. Below are some considerations associated with both approaches.

4.3.1 Quantitative Exposure and Risk Assessment

The quantitative approach to determining health risk uses data collected through exposure monitoring, ventilation or air dispersion modeling, process monitoring, and area monitoring to identify contaminants and concentrations of concern. Example data that is used as part of the quantitative approach is shown in Figure 4.2.

Often, the use of quantitative methods is challenging since many times the necessary data is not available to complete the assessments, or if available, it may only be available for some of the contaminants, which may pose an increased health risk to the worker.

During the early years of and prior to the existence of the Occupational Safety and Health Administration (OSHA), workplace monitoring through data collection was not commonly performed in the United States. As the industrial hygiene profession has matured, it has become an expectation to collect workplace data, such as

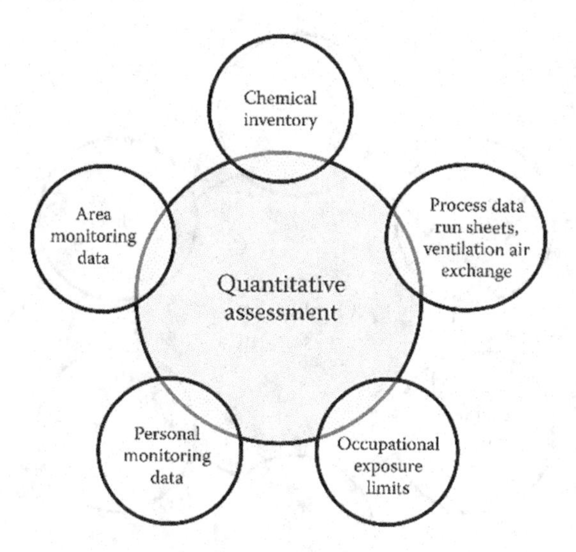

FIGURE 4.2 Components of quantitative assessment.

personal or area monitoring data, to characterize the work environment and actual worker exposure to various workplace constituents. In addition, today's industrial hygienist is also using process monitoring data, collectively with other sampling and monitoring data, to gain a complete picture of the occupational health risks to the worker.

Health risks may arise from a single contaminant or can be from several contaminants which together represent an even greater health risk (synergistic effects). With the development and passage of the OSHA regulations, data collection is now a standard step in the risk assessment process, since in many cases the regulations require collection of workplace data.

In the future, occupational safety and health regulations related to health risks posed in the workplace are anticipated to increase because of a growing public desire to live a healthier, longer life, in a controlled and responsible manner. Many practicing industrial hygienists believe that more sampling and monitoring data, even negative exposure data and conservative decision-making, will be needed to minimize future liability and worker compensation settlements.

4.3.2 Qualitative Exposure and Risk Assessment

When conducting exposure or risk assessments, there are times when data cannot be obtained to quantify concentrations of contaminants. For example, it may not be possible to sample the air stream because it may be too radioactive. Therefore, the need arises to conduct a risk assessment using qualitative data only. A qualitative risk assessment is based on the use of information, such as process-based, observed workplace conditions, and professional judgment of the industrial hygienist. Figure 4.3

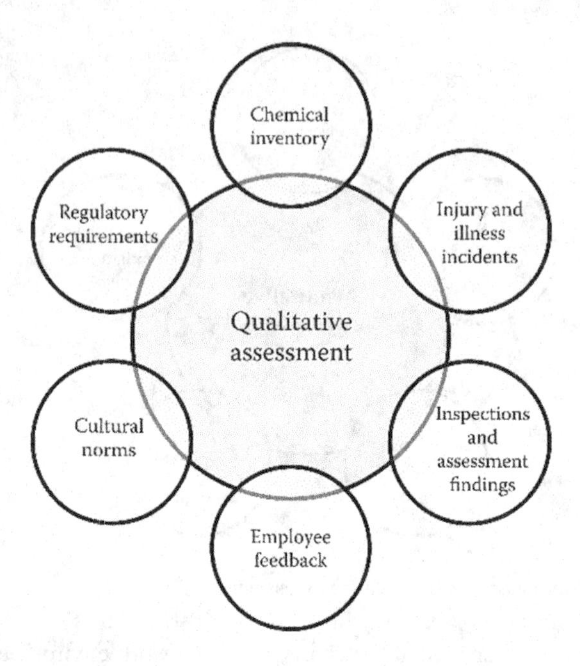

FIGURE 4.3 Components of qualitative risk assessment.

lists common information that is used in conducting a comprehensive qualitative assessment.

If the industrial hygienist uses a qualitative approach to characterizing and monitoring exposures, then conservative decision making must be applied (i.e., safety factors) to consider uncertainties in the risk identification and evaluation process.

4.4 PROCESS FLOW OF EXPOSURE ASSESSMENT

The industrial hygienist may use either a quantitative or qualitative approach, or a combination of both, to implement the exposure assessment process; however, the industrial hygienist generally performs at least nine steps to fully implement the exposure assessment process. Figure 4.4 depicts the nine basic steps of the exposure assessment process model. In order for the exposure assessment process to be deemed valuable and effective, a well-thought-out process must be designed and implemented.

Attributes of a well-designed exposure assessment and monitoring process must include some key elements, such as:

1. Developing an exposure monitoring strategy complementary to the constituent being monitored. The exposure strategy should target contaminants of concern and contaminants which may or may not have established occupational exposure limit.

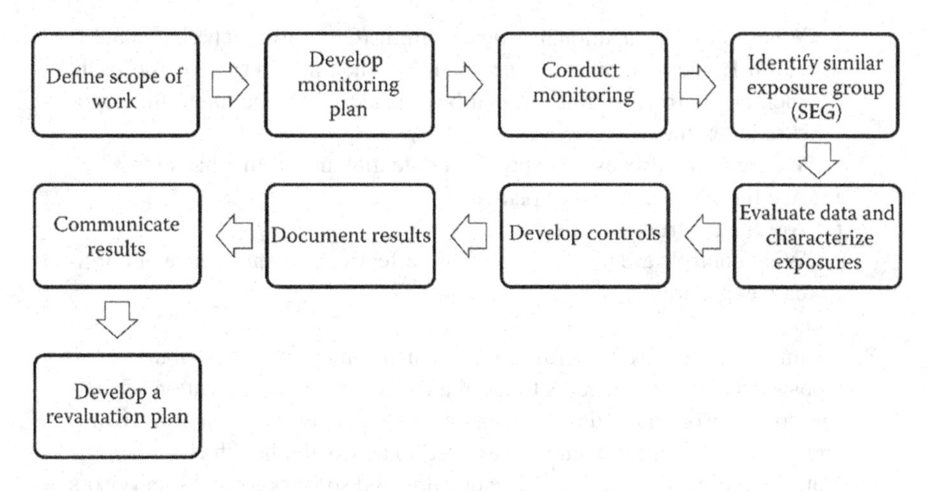

FIGURE 4.4 Exposure monitoring process model.

2. Ensuring the selection of an analytical method that is acceptable by an independent, credible organization (e.g., the American Society for Testing and Materials [ASTM] and the National Institute for Occupational Safety and Health [NIOSH]).
3. Developing, implementing, and documenting quality control procedures. The use of quality control procedures ensures the data collected is quantifiably representative of workplace conditions.
4. Developing monitoring plan. The monitoring plan is used to determine which employees and work evolutions require data to determine and quantify exposures. The main objectives of a monitoring plan are to quantify exposures and monitor if the controls in place are controlling exposures.
5. Conducting monitoring. When conducting exposure monitoring, it is important that the strategy be implemented as written using the appropriate sampling media and placing the monitoring equipment in the best location and position to represent actual exposures to workers.
6. Identifying similar exposure groups (SEGs). An SEG is a group of employees who have the same or similar exposure to the same chemical agents due to the tasks they perform. Considerations must be given to the tasks, process, procedures, and practices used in preforming the tasks, chemicals and materials used, and time it takes to perform the task.
7. Assessing and characterizing exposures. This step may be considered one of the most important in the process. To determine exposures, some fundamental questions must be explored, and some of these are listed below:
   a. What are the chemical, physical, and biological substances in the work environment?
   b. What are the potential health effects associated with overexposure? Also, keep in mind that some chemicals may create synergistic or compounding effects when exposures are combined.

c. What is the occupational exposure limit (OEL) for each chemical? If an OEL does not exist for the chemical, then an alternative approach, such as control banding, should be considered to document the exposure assessment.
d. Where in the process is exposure a potential and from what source?
e. Are there any controls in place?
f. Are the controls effective?
g. What controls are needed to increase effectiveness and reduce or eliminate exposures?

8. Documenting results. Recording the results is an important attribute of the exposure assessment process to facilitate effective communication of findings so that workers and supervisors can adequately understand the exposure profile. If control banding was used to assess the health risk, then the control banding analysis should be documented so workers and supervisors can understand how they are being protected and health risk minimized.
9. Communicating results. Communication of exposures or lack thereof must be done with care and sensitivity.
10. Developing controls. Controls are developed based on the exposure assessment results. Controls should take into account each potential route of exposure (inhalation, ingestion, injection, and absorption).
11. Developing and implementing a plan and making a schedule to reevaluate. The exposure assessment process should be repeated to verify that the control methods instituted improve exposure potential and to ensure that exposures to workers are well below the OEL or determined health risk.

These key elements are discussed in further detail as the exposure monitoring process model is elaborated below in Sections 4.4.1–4.4.10.

4.4.1 Defining the Scope of Work

Defining the scope of work to be performed by workers can appear on the surface to be an easy task. For example, if the work to be performed consists of changing out a hose on a tank, some people would define the scope of work in generic terms:

- Identify the hose to replace
- Make sure you have the parts in the warehouse
- Change the hose
- Dispose of all waste generated as part of the job

Although workers can perform the work, there are a number of steps that must be performed, in a certain sequence, in order to appropriately identify the hazards associated with each work step. It is up to the industrial hygienist and workers to understand all steps to be performed, and the hazards associated with each step.

Often, companies will use a collaborative effort, between subject matter experts and workers who have knowledge of the task to define the work steps. The hazard identification step must factor in all the resources needed to accomplish the work, along with understanding hazards that could impact each worker. Once the work scope has been identified, then similar exposure work groups can be developed and/or assigned.

4.4.2 Developing a Monitoring Plan

Fundamental to the exposure assessment process is the collection of characterization data that will be used in assessing a worker's exposure and health risk from workplace hazards. Questions that are directly relevant to the characterization process include:

- What are the work processes being performed in the manufacturing or execution of the company business?
- Have these processes been in place for a long time, or have improvements to the process been executed?
- What personnel are assigned to each task associated with performing work?
- What physical, chemical, biological, or radiological hazards are associated with the work processes?
- Have procedures been established that document the process for performing workplace characterizations, including understanding the limitations of the equipment being used to monitor such exposures?

To gain an understanding of what characterization data is available, the industrial hygienist should review historical data, along with collecting any additional data that could supplement any data gaps. Historical characterization data can be found from a number of resources. Table 4.1 contains an example of documents the industrial hygienist should consider when collecting characterization data. Please note that this list should be tailored to the type of manufacturing or business operations being performed and is used in both a qualitative or quantitative approach to monitoring. Review of the records should not only identify the data generated, but also consider the instrumentation used for collecting the data, along with confirming the calibration of the instrumentation. The industrial hygienist should understand whether the data was collected over a specified time period (i.e., every 5 minutes or over an 8-hour time frame), was associated with area or personal monitoring, and if there were any process changes that occurred since the data was collected. If relevant, environmental conditions occurring at the time of collecting the data can also be useful. In addition, the industrial hygienist needs to understand the physical properties associated with the contaminant, such as whether the contaminant behaves as a solid, liquid, vapor, or gas.

Most companies have an ongoing workplace monitoring program, so the industrial hygienist should also evaluate current data collection processes, which would include understanding sampling practices and procedures, data collection methods,

TABLE 4.1
Example of Resource Documents for Characterization Data

Employee sampling of database records
Historical work planning documents
Operational work packages
Operational process monitoring reports
Operational rounds and routine reports
Maintenance work packages
Chemical inventories and monitoring (if available)
Environmental permits and regulatory reports
Post-job review reports
Worker injury reports
Workers' compensation records
Employee job task analysis reports
Historical regulatory actions

and current instrumentation calibration practices. Current sampling and monitoring data, along with historical data, will form the basis of characterization data that will be used in the exposure assessment process. As input into the development of an exposure assessment strategy for workers, it is important for the industrial hygienist to understand the health risks posed by the exposure being monitored and controlled.

The traditional data characterization and exposure assessment model used by industrial hygiene is generally based on toxicological studies for carcinogenic and noncarcinogenic contaminants. Characterization data is collected, evaluated, and applied in the exposure assessment and hazard control process in the same manner; however, depending on the contaminants, the exposure model applied can vary depending on whether the contaminant exhibits an acute and/or chronic health hazard. For example, one carcinogen, radiation, applies the linear no-threshold (LNT) model in the hazard analysis and control process.

The characterization and monitoring plan is usually driven by one of the below two purposes:

- To identify baseline contaminants of concern related to worker exposure
- The ongoing monitoring needed to address either changing environmental conditions of changing contaminants (e.g., changing chemical products in the workplace), or because the work steps of personnel performing work have changed

The characterization and monitoring plan generally addresses the following:

- Work location
- Work activity

- Potential contaminants of concern (contaminants identified through historical and current operational research)
- Sampling type (personal or area)
- Sampling frequency (continuous, four or eight hours, or every shift)
- Sampling instrument (sorbent tube or personal sampling pump)

Figure 4.5 is an example of a typical characterization and monitoring plan.

4.4.3 Implementing the Characterization and Monitoring Plan

Once the scope of work has been identified, and a characterization and monitoring plan has been developed, it is up to the industrial hygienist to implement the plan. There are two primary methods used to collect data:

Process Description: Remove legacy asbestos insulation—friable

Process Location: Building 572, Room 158

Sampling type personal or area: Both personal and area samples are to be collected.

- Area samples must be located at the entrance to the containment and at the exhaust of the air moving equipment.
- Personal samples must be located in the breathing zone of each worker.
- Conduct five area clearance samples at the completion of the job using forced air distribution.

SEGs:

- SEG 1: Asbestos maintenance mechanic, asbestos inspector
- SEG 2: Asbestos laborers, industrial hygienist

Sampling collection:

- Recommended sampling rate: 0.5–5.0 L/min
- Recommended air volumes: Minimum: 25 L Maximum: 2400 L
- Sampling media: 25 mm diameter cassette containing a mixed-cellulose ester filter equipped with an electrically conductive 50 mm extension cowl

Laboratory analytical method: Transmission electron microscopy (TEM)—NIOSH Method 7402

- Sensitivity of the method
- Accuracy of the method
- Analysis time
- Cost
- Availability of equipment

FIGURE 4.5 Sampling plan for asbestos.

- Data that results from collecting samples on a media that is analyzed by an accredited laboratory. These samples may be obtained by means such as a gas sampling bag, filter, sorbent tube, or wipe method. The results of these types of samples are not available at the time the sample is collected, and the lag time between sampling and receiving the data can range from hours to days.
- Collection of data using a direct reading instrument. In such cases, the data is available at the time the sample is taken, with no delay.

Collecting air samples to be analyzed for a particular constituent can oftentimes present some challenges if the appropriate laboratory analysis method is not selected and the sample is not collected with the desired analytical method in mind. A complete list of sampling and analysis procedures for a host of chemicals can be found at the OSHA website at www.osha.gov/dts/sltc/methods/inorganic/id160/id160.html or the NIOSH manual for analytical methods at http://www.cdc.gov/niosh/docs/2003-154/default.html for companies doing business in the United States.

When preparing to collect samples to quantify exposures, it is pertinent that the sample analytical method be known, along with understanding what is the minimum sample mass need, as well as the sample collection packaging and storage methods.

4.4.4 Similar Exposure Groups

The term *similar exposure groups* (SEG) refer to grouping or organizing employees into monitoring groups based on similar work assignments and similar contaminant exposure profiles. Often, the groups are organized by the similarity in the types of work performed, frequency with which they are performed, material and equipment being used, and amount of anticipated exposure to a particular contaminant. Table 4.2 provides an example of four SEGs.

The number of SEGs that are identified for a specific task or project is dependent on:

1. The number of tasks being performed and the differing types and levels of exposure.
2. The potential for the employee to be exposed to contaminant concentrations that may cause acute and/or chronic health impacts.
3. The industry and expected level of documentation needed to provide an adequate level of protectiveness.

In some industries, such as the government industries, there are special laws, such as the Energy Employees Occupational Illness Compensation Program (EEOICP), that provide additional compensation above the traditional workers' compensation costs to employees who can prove that an acute or chronic illness was caused by a potential exposure to chemicals and radiation. When the identification of SEGs has been completed, it is necessary to evaluate the data collected to determine adequacy of the strategy to reduce risk resulting from performing work.

TABLE 4.2
Similar Exposure Group Example

SEG Category	Job Title	Work Responsibility
1	Pipe fitter	Configure and install pipes, support hangers, and hydraulic cylinders; cut, weld, and thread pipes in the shop and on the job; remove water from flooded areas, such as manholes and other confined spaces
2	Housekeeping	Remove trash and clean containers; sweep, scrub, mop, and polish floors; clean carpets, rugs, and draperies with a vacuum cleaner; dust furniture and clean all fixtures; wash windows
3	Mechanical engineer	Design and implement safe and reliable processes and systems; develop and test theoretical designs and applications; monitor and evaluate the performance of processes and systems; serve as subject matter expert on all plant mechanical systems and equipment
4	Chemist	Analyze compounds to determine chemical and physical properties, composition, structure, relationships, and reactions, utilizing chromatography, spectroscopy, and spectrophotometry techniques to develop standards and prepare solutions and reagents for testing

4.4.5 Occupational Exposure Control Banding

In the United States, the National Institute for Occupational Safety and Health (NIOSH) has developed an occupational exposure banding (OEB) process as a systematic approach that uses both qualitative and quantitative hazard information on selected health-effect endpoints to identify potential exposure ranges or categories.[3] Benefits of applying the NIOSH OEB process include:

- A three-tiered system that allows industrial hygienists of varying expertise to use the process,
- Determination of potential health impacts based on nine health endpoints,
- Hazard-based categories linked to quantitative exposure ranges, and
- Assessment of the process through extensive evaluation to determine consistency of the OEB process with OELs.

Each tier of the process has different requirements for data sufficiency, which allows a variety of people to use the process in different situations. The most appropriate tier for banding depends on the availability and quality of the data, how it will be used, and the training and experience of the industrial hygienist. It should be noted that the OEB is not meant to replace an OEL; rather, the OEB serves as a starting point to inform risk management decisions when OEL is not available. Industrial hygienists can band a chemical manually, or by using the occupational exposure banding e-Tool from NIOSH.

4.4.6 Evaluating Data and Characterizing Exposures

Evaluating data and understanding and characterizing exposures resulting from conducting employee exposure monitoring is key in assessing and reducing the risk of exposure to various chemical and biological constituents. Items to consider when evaluating the data should include.

- Exposed population
- Task being performed
- Toxicity and health risk (e.g., carcinogen vs. noncarcinogen)
- Occupational exposure limit (if available) or data gathered as part of OEB
- Potential sensitization of worker population (e.g., latex gloves)
- Exposure time
- Additive and synergistic effects of contaminants

4.4.6.1 Application of Linear No-Threshold Versus Linear Threshold Models

Chemicals (both organic and inorganic) and other contaminants (i.e., biological and radiological) have been studied and modeled over the years to determine toxicological properties, including whether a contaminant will pose an acute or chronic health hazard. Generally, this information can be found in the NIOSH pocket guide or other industrial hygiene information resources. An example of an acute health hazard would be exposure to ammonia. Although the chemical is classified as an irritant, and can be caustic if overexposed to large quantities, it is predominantly managed as an acute hazard. Conversely, exposure to beryllium may be considered an acute health hazard because it causes sensitization, but it is also considered a chronic health hazard because medically any exposure to beryllium can potentially cause sensitization and accumulate over time in the body and lead to chronic beryllium disease. Radiation is also known to cause acute health effects if one is exposed to a large dose, but predominantly, radiation and radioactivity are controlled as chronic health hazards.

Ionizing radiation is a Group 1 carcinogen as defined by the International Agency for Research on Cancer (IARC). By definition, Group 1 carcinogens are contaminants that have shown sufficient evidence of carcinogenicity in humans. In addition to ionizing radiation, other examples of Group 1 carcinogens include asbestos, benzene, cadmium, and polychlorinated biphenyls (PCBs). Predominantly, the exposure model applied to ionizing radiation, and still in existence today for many carcinogens, is the LNT model.

The fundamental principle of the LNT model is based on the premise that any level of exposure to radiation causes some level of harm. Even though the level of harm may be small and minute, over time the effect is cumulative and exposure to radiation should be kept as low as reasonably achievable (ALARA). This exposure model and theory was largely adopted and gained momentum in 1956 by the U.S. National Academy of Sciences Committee on Biological Effects of Atomic Radiation (BEAR I) in 1956.

The radiation hormesis model provides that exposure of the human body to low levels of ionizing radiation is beneficial and protects the human body against deleterious effects of high levels of radiation. The LNT model, on the other hand, provides that radiation is always considered harmful, there is no safety threshold, and biological damage caused by ionizing radiation (essentially the cancer risk) is directly proportional to the amount of radiation exposure to the human body (response linearity). Chronic health impacts from exposure to ionizing radiation are directly proportional to dose. Over the years, there has been debate about whether low levels of radiation can cause positive health benefits. The LNT model is still the most recognized exposure model in health physics and forms the basis of regulatory requirements. Chemicals, such as beryllium or cancer-causing chemicals, may be managed in a fashion similar to radioactivity and the LNT model. Predominately, the linear threshold (LT) model is widely used in industrial hygiene and the exposure assessment process for noncarcinogenic contaminants.

The LT exposure model is postulated based on the premise that there is a threshold, for a particular contaminant, that acute and chronic health impacts can be managed against to prevent employee harm. Industrial hygienists most commonly apply the LT model in the workplace based on evaluating characterization and monitoring results against a regulatory value to determine whether the workplace hazard is acceptable. It is up to the industrial hygienist to determine which exposure model will be applied when evaluating characterization and monitoring data; however, whatever exposure model is applied it must be defensible. The results of that analysis will then be used to support any future occupational illness claims, and to support future work planning efforts. When determining the exposure model to apply when evaluating data, the industrial hygienist should consider the following:

- Cancer and some disease exposure monitoring is driven by compliance monitoring because of the liability of acute or chronic diseases.
- Contaminants (e.g., radioactivity or cancer-causing chemicals) that follow the LNT model for risk apply the ALARA principle and are compliance monitoring driven.
- The LT model and comprehensive exposure assessments characterize all exposure for all workers for all days.
- Reliance on personal protective equipment versus data collection and managing hazards by engineering and administrative controls.

4.4.6.2 Occupational Exposure and Administrative Control Limits

OELs have been set for many chemicals by several organizations. Some of those agencies are highlighted in this section. Although there are several agencies that set OELs for various chemical constituents, the permissible exposure limits set by OSHA are the only OELs that are widely considered legally enforceable. However, other limits can be enforceable if they are included as a contractual requirement or if the OEL has been written in the company's procedures or business plan.

The AIHA recognized that in the absence of regulatory or contractual OELs, the industrial hygienist may establish "working OELs" to differentiate promulgated and non-promulgated OELs. Working OELs can be established in a number of ways; however, OEB is a very effective approach to establishing an OEL which is primarily toxicologically based. Working OELs are established in the absence of formal OELs, or they may be established in the presence of a formal OEL when there is significant uncertainty about the adequacy of an established or formal OEL (Mulhausen and Damiano, 1998).[4] The OEL agencies that are commonly used in comparing exposures in the United States are shown in Table 4.3.

Often, the OELs published by the primary agencies are the same, and in some instances, they can be quite different depending on how often the agencies revise their exposure limits. It is important before beginning a sampling campaign to consider the various OELs and determine the OEL that is applicable to the exposure and enforcement profile of your company. Some companies have adopted the position that the most restrictive OEL should be used to further ensure that workers are protected from adverse impacts of hazards in the workplace and health risks are reduced. This philosophy is not the norm because it can introduce additional costs to the implementation of a safety and health program.

The two most widely used OELs are published by OSHA and the American Conference for Government Industrial Hygienist (ACGIH). A comparison of the OELs published by agencies is listed in Table 4.4 for some chemicals that are frequently found in the workplace. The table is populated with some OELs to make the point that OELs published by different agencies may very well be the same, but they may also be different. The decision must be made as to which exposure limit is applicable to the company exposure profile.

TABLE 4.3
Occupational Exposure Agencies

OEL Agency	OEL
Occupational Safety and Health Administration (OSHA)	Permissible exposure limit (PEL)
American Conference for Government Industrial Hygienist (ACGIH)	Threshold limit value (TLV)
American Conference for Government Industrial Hygienist (ACGIH)	Biological exposure indices (BEIs)
National Institute for Occupational Safety and Health (NIOSH)	Recommended exposure limit (REL)
California (State) Occupational Safety and Health Administration (CalOSHA)	Cal (state-dependent)/OSHA/PEL

TABLE 4.4
Comparison of OEL

Chemical	OSHA (PEL)	ACGIH (TLV)	NIOSH (REL)
Ammonia	TWA 50 ppm	TWA 25 ppm	TWA 25 ppm
Arsenic	TWA 0.5 mg/m^3	TWA 0.01 mg/m^3	None
Chlorine	C 1 ppm	TWA 0.5 ppm	C 0.5 ppm
Chromium	TWA 0.5 mg/m^3	TWA 0.5 mg/m^3	TWA 0.5 mg/m^3
Fluorine	TWA 0.1 ppm	TWA 1 ppm	TWA 0.1 ppm
Iron (oxide dust and fume)	TWA 10 mg/m^3	TWA 5 mg/m^3	TWA 5 mg/m^3
Lead	TWA 0.050 mg/m^3	TWA 0.05 mg/m^3	TWA 0.050 mg/m^3
Silica	TWA 20 mppcf (80 mg/m^3/%SiO_2)	TWA 0.025 mg/m^3	TWA 6 mg/m^3
Sulfuric acid	TWA 1 mg/m^3	TWA 0.02 mg/m^3	TWA 1 mg/m^3
Toluene	TWA 200 (C), 300, and 500 ppm	TWA 20 ppm	TWA 100 ppm
Xylene	TWA 100 ppm	TWA 100 ppm	TWA 100 ppm

Note: C, ceiling value; TWA, time-weighted average; mppcf, million particles per cubic foot.

4.4.7 Develop Controls

Specific controls should be developed and implemented to reduce or eliminate hazards that have been identified by the industrial hygienist, worker, or management. Implementation of controls takes into consideration the hazards and potential exposure risks. Workplace hazard controls take on many forms, such as implementing engineering measures, substituting chemicals, changing the process or work practices, or even relocating the worker.

Further detail on the development and application of hazard controls is discussed in Chapter 6; however, whichever method is selected, the control method must be effective in risk reduction and exposure mitigation. When developing hazard controls, the following should be kept in the forefront of the decision-making and selection process:

- The hazard control must be selected based on the traditional hierarchy of controls, as required by OSHA: product substitution, engineering controls, administrative controls, and personal protective equipment.
- Select controls that are valid and tested. Do not select controls that are based on hypothetical or old data, assumptions, and unproven theories unless another layer of protection is added.
- Ensure that the controls selected do not introduce additional hazards into the workspace, for example, using ventilation to control vapors that introduces a high noise level into the work environment.
- Educate workers on the control methods and the purpose of implementation.

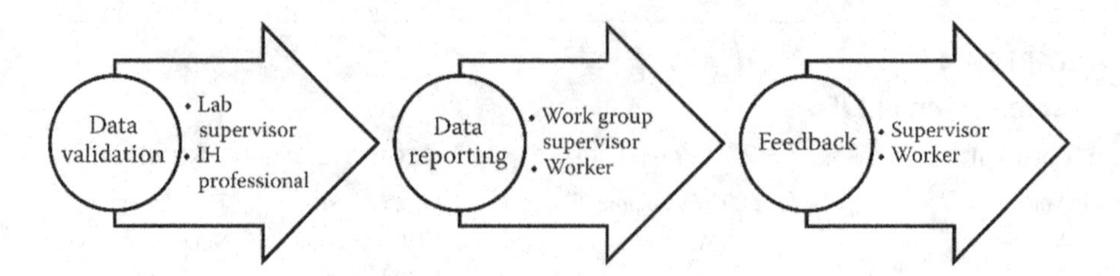

FIGURE 4.6 Example of data handling process. IH, industrial hygiene.

4.4.8 Document Results

The documentation of analytical results includes not only thoroughly documenting the methods applied to collect and summarize the data, but also validating the data. The data handling process is shown in Figure 4.6. Before reporting data, thorough validation of the data is required. Reporting data that has not been validated is risky since there is always the possibility of data error, which can result from improper collection methods, error in the laboratory analysis method, or error in the annotated results. Such errors can lead to incorrect reporting of potential exposures to workers, as well as loss of trust from workers if errors are later revealed.

4.4.8.1 Peer Review and Validation

The peer review process is integral to scholarly writings and is used extensively in the research arena. This form of fact checking also has applicability in other areas and can provide credibility and validity to the result of an assessment, as well as the process used to determine decisions. Peer review is designed to prevent the publishing of inaccurate findings, unacceptable interpretations, and personal assumptions and views. It relies on colleagues having knowledge in the area in which they have been asked to provide a review, and making an informed decision about whether the assessment and the results represent a comprehensive assessment. Peer review is a great way to scrutinize and validate important attributes, such as:

- Assumptions
- Calculations
- Interpretations
- Methodology used during the process
- Acceptance criteria for risk

Using peer review to validate risk assessments is a good way to add credibility to the results when presented to others. It also demonstrates that the decision is being viewed as solid and defensible by those involved in the conclusion and the decision-making process.

Once the workplace exposure monitoring campaign has been completed and samples have been analyzed, the results should be compared with the OELs. These results may shed light on

- Whether employees are exposed to workplace airborne hazards
- Whether the current methods of control are providing adequate protection
- Whether improvements are needed in the control methods being used

An example of a typical exposure monitoring reporting form is shown in Figure 4.7.

4.4.9 Communicate Data Results

Communicating exposure-related information to workers and management is just as critical as collecting and analyzing the exposure data. It is recognized that most

Name:________________ Location: ________________

Contaminant: ________________

Summary of task monitored:

Results: ________________ OEL: ________________

Summary of results:

Sample collected by:

Signature: ________________

Results reviewed and validated by: ________________

FIGURE 4.7 Example of personal monitoring exposure result.

industrial hygiene professionals are skilled in collecting, analyzing, and interpreting the data; however, many are lacking in the skills of effective communication. In such cases, this presents a challenge for the industrial hygienist when communicating exposure data to the workers, especially when the data indicates that an exposure has occurred. The industrial hygienist has more than one method that can be used to communicate information. It is important to be able to ascertain when each type of communication will be appropriate for the various situations encountered. Communication of data can occur in person, via e-mail, or through formal documented means. When communication does not take place in an open and honest manner with clarity, a lack of trust and credibility often results between the industrial hygienist and the workforce.

4.4.10 Develop a Re-Evaluation Plan

Conducting an exposure assessment or a risk characterization is not a one-time event or activity. The mistake that some industrial hygienists make is to conduct an initial evaluation and fail to perform a re-evaluation to periodically validate workplace conditions. A typical re-evaluation plan is designed with the same elements that one would expect in an initial assessment strategy, with the exception of the inclusion of all data and information obtained since the last assessment or evaluation was performed.

The re-evaluation plan may be driven by the need to gather initial data to validate the hazard's baseline, or to verify that environmental and workplace conditions have not changed since the last hazard assessment. In addition, a re-evaluation plan should be considered to validate and make additional improvements to the existing monitoring strategy. There are a number of reasons why the industrial hygienist would need to perform a re-evaluation; however, the important point to remember is that re-evaluation of workplace conditions is necessary to improve the hazard analysis and control process.

4.5 OCCUPATIONAL SAFETY AND HEALTH CHARACTERIZATION AND MONITORING EQUIPMENT

Exposure monitoring equipment is a critical and essential element in the process to characterize exposures. Monitoring equipment is used to help characterize and evaluate work areas where there is a potential for a hazardous environment to exist. Environmental conditions can be monitored in a variety of ways. When collecting samples to quantify worker exposures, the following equipments are typically used:

- Diffusion detector tubes
- Vapor monitoring badges
- Personal air sampling pumps

If the desire is to sample specific areas, then consideration should be given to the use of the following equipments as an option to collect data:

- Detector tubes
- Sampling pumps
- Handheld electronic monitors
- Fixed wall-mounted electronic monitors

When implementing the sampling strategy plan, careful consideration must be given to the type of equipment needed to collect the data. A brief summary of the utility of each type of sampling equipment is discussed below.

4.5.1 Diffusion Detector Tubes

Gas diffusion detector tubes are easy to use when sampling for gases and vapors. They also offer a low-cost method when collecting a large number of samples. These tubes are accurate enough to provide an indication of the contaminant concentration in the workplace. A detector tube is a graduated glass tube filled with a chemical reagent that will produce a color change when exposed to the gas in question. It is used with a hand pump that will draw a known amount of sample into the tube.

There are two main types of pumps available to use in this system: a piston pump or a bellows pump. The piston-style pump requires the user to pull a piston to pull air through the tube. With the bellows style, the user squeezes the bellows, and upon release, air is pulled through the tube as the bellows opens. The tubes generally are sealed at both ends and ready to use when the tips are broken off. Gas detector tubes are available for hundreds of compounds and have been used for many years. A major advantage in using this system is that it allows for sampling of an area quickly, with minimal expense. Although the cost associated with the use of these tubes is low and these tubes are easy to use, there are some disadvantages of using them, such as:

- The potential for chemical cross-sensitivity and interference with other chemicals.
- Detector tubes are not the most accurate method for collecting air samples and quantifying exposures. The accuracy assigned to these tubes falls in the area of approximately ±20%.
- Detector tubes have a shelf life, and many of them have to be stored in refrigerated environments. Many of the tubes are temperature sensitive.
- Many of these tubes require that multiple pump strokes be taken to achieve the desired sensitivity.
- Errors can occur from failure to complete the necessary amount of pump strokes or from not allowing the time needed for each sample to work its way through the tubes prior to activating another pump stroke.
- Some chemicals can interfere with the tube and give false-positive readings.

To effectively use these tubes, the following steps are necessary:

- One end is broken off and the tube is placed in a tube holder. If the tube has a clip, it can be hung on the worker's lapel, near the breathing zone, to get an accurate reading of the worker's exposure. Be cautious, as the end of the glass tube has been broken and if not cut smoothly, the worker may sustain cuts by coming into contact with the tube end, which can cause injury.
- To ensure that the accurate exposure level is calculated, record the sampling start time on the writing area of the tube. The indicated area has a calibrated scale. This is where the reaction with the sample gas or vapor takes place, causing a discoloration that can be read on the scale.
- At the end of the work shift, remove the tube from the worker and record the removal time.

Exposure can be determined by observing the discoloration and recording the parts per million (ppm) value from the scale. The time-weighted average is obtained by dividing the scale value by the total amount of time sampled (in hours).

4.5.2 Vapor Monitor Badges

Vapor monitor badges work through the concept of diffusion and are an excellent option that can be used to monitor a worker's breathing zone to determine exposures. These badges are typically used to determine an 8-hour time-weighted average or a 15-minute short-term exposure limit. The exposure results obtained from vapor monitor badges are more accurate that those obtained from diffusion tubes. However, these badges have several disadvantages, such as:

- They must be sent to a laboratory for analysis.
- The results are not immediately available.
- These badges are only available for a limited number of chemicals.
- They are oftentimes more expensive than diffusion tubes.

Vapor monitor badges are available for compounds such as formaldehyde, organic vapors, ethylene oxide, mercury, and nitrous oxide. These badges are lightweight and clip onto the worker's collar. In order to correctly calculate and report exposures at the end of the work shift, the following information must be collected:

- Exposure time
- Temperature
- Relative humidity
- Date exposed
- Employee identification data
- Vapor monitor badge number

Vapor monitor badges are frequently used as a monitoring option because they are easy to use; all that is needed is to have the worker wear the badge and submit it to the laboratory to be analyzed. Vapor monitor badges are usually analyzed by a laboratory by desorbing the vapors trapped on the badge. The desorbed vapor is then run through a gas chromatograph to determine the contaminant level.

4.5.3 Personal Air Sampling Pumps

Personal air sampling pumps are a little more difficult to use than vapor monitor badges and detector tubes. However, the results are more accurate than those of other methods and allows for the sampling of many more different types of chemicals. On the other hand, there are some distinct disadvantages of using sampling pumps, such as:

- They require familiarization and knowledge of laboratory collection methods, such as NIOSH sampling methods.
- The type of collection media used is specific to the laboratory analysis method.
- They are bulkier and more cumbersome than other sampling methods.
- Because of the bulky construction, workers are not always willing to wear the pumps for an entire work shift.

The air sampling pump assembly and sampling train consist of the following:

- A pump that pulls a constant amount of air through a charcoal tube or a filter cassette.
- A tube to connect the pump to the charcoal tube or filter cassette.
- A clip or method for attaching the collection media near the worker breathing zone.

These pumps require calibration before each use and must have the flow rate set to the appropriate level based on the chemical constituent being monitored. The worker is expected to wear the pump with the collection media attached for the entire work shift or until the task or exposure period ends for the workday. Once the work shift is finished, the charcoal tube or filter cassette is then sent to the laboratory for analysis.

4.5.4 Handheld Electronic Monitors

Handheld monitors can range from simple single-gas monitors to somewhat complex four-gas monitors with data logging capability. These monitors have distinct advantages and disadvantages, which are listed below. Disadvantages include:

- Handheld electronic monitors can experience interference that can alter the reading from gases that are similar in components.

- The operator must have adequate knowledge to be able to interpret the data obtained and must also be familiar with the calibration and limitations of the device.
- In comparison with other types of monitors, they are often more expensive.

Advantages include:

- The readout is instantaneous and in real time.
- There is no waiting for a lab to analyze the results.
- The result displayed is the concentration at that moment in time.
- Many handheld meters have a visible and/or audible alarm to alert the user if the concentration is above a safe level.

4.5.5 Fixed Air Monitors

Fixed air monitors are similar to handheld monitors, except that they are mounted in a specified area and do not require an operator on location. Fixed monitors use sensors similar to those of handheld monitors and often do not have a wide range of chemicals that they can detect. These monitors are convenient because they operate at all times and often have audible and/or visual alarms to alert workers of dangerous environments. Fixed air monitors are used in locations to warn personnel of approaching administrative exposure levels (50% of the occupational exposure level). Fixed monitors have advantages and disadvantages similar to those of handheld monitors.

4.6 CASE STUDIES TO FACILITATE THOUGHTFUL LEARNING

Below are several case studies which represent situations that the industrial hygienist routinely encounters. The purpose of the case studies is to simulate thought as to how an industrial hygienist would respond and communicate the information to workers and management.

4.6.1 The Presence of an Intermittent Odor

Oftentimes an industrial hygienist will be informed by a worker that he/she was working in an area and "smelled something" unusual and unrecognizable. Generally, she will also state that she has no idea what is the source of the smell. In such cases, the industrial hygienist would collect as much information from the worker as possible and proceed to investigate the issue.

In this case, you, as the industrial hygienist, conducted a thorough interview of the worker and documented her account of what she believed had occurred in the workplace. She further stated that the smell was noticed intermittently and generally did not persist for long periods of time.

You visited the work area and conducted a visual walk-through survey. During the walk-through, you did not notice any unusual odor; therefore, you concluded that the work area was safe, and no further action was needed. The workers re-entered the work area and proceeded to work. After about two hours, the initial worker once again noticed the presence of the noxious odor, exited the workplace, and contacted her supervisor.

The supervisor contacted you once again and insisted that a thorough hazard review of the project be conducted. You solicited assistance from colleagues and your manager in conducting a full-blown review of the project, with assistance from workers and their supervisor. The team discovered that at certain stages in the process, a short burst of ammonia was being emitted into collected samples that revealed that the level of ammonia being emitted exceeded the OEL for ammonia. The sampling results were communicated to the work group supervisor and the work group, and appropriate controls were put in place to control the potential overexposure. In addition, information was provided to the medical provider to determine if additional medical testing was needed and if the worker should be included in a medical surveillance program.

4.6.2 The Presence of an Intermittent Odor Lessons Learned

Some of the lessons that can be gleaned from the case study are listed below:

1. At the initial notification by the worker of an unusual odor being detected, you should have completed a thorough investigation that included activities such as:
   a. Requesting the supervisor to remove workers from the work area until a hazard evaluation could be completed.
   b. Identifying what task was being performed when the odor was noticed.
   c. Determining what chemicals were used in the process and conducting a full-shift monitoring campaign to determine if any of the chemicals were used in quantities that could be harmful to workers, or that the industrial process, which required the use of chemicals, had been thoroughly evaluated and hazards identified.
2. You did not initially perform an adequate hazard analysis before sending the workers back to work.
3. Oftentimes, in order to get to the root of an issue, a team of people that includes the workers is necessary because the workers are generally the most knowledgeable in the tasks they perform daily.

4.6.3 "I Have Been Sick for the Past 6 Months"

Although industrial hygienists are focused on the prevention of injuries and illnesses, they cannot be in the workplace all the time. There are situations where workers may present themselves as possibly having an illness that could be related to exposure to contaminants at work. For example, a worker approaches you, the industrial hygienist, and states that he has not felt well for the past several months. This worker has approached you on more than one occasion, and the last time the worker discussed his health, he claimed that he had been diagnosed with emphysema but has never smoked. The company you both are employed by is related to the coal mining industry.

As an industrial hygienist, you have routinely monitored the worker's breathing zone and not identified any overexposures to coal and coal waste products; however, there have been recent changes to the handling of waste products generated from the processing facility. As an industrial hygienist, you start to wonder if, in fact, there could be a connection between work activities and potential overexposure of workers to toxic contaminants.

You proceed to investigate the worker's daily tasks and patterns of performing work. Additional monitoring data is collected, and additional analytical tests are performed by the laboratory. Also, you interview other workers performing those same tasks, along with engineers who were involved in the recent changes in waste handling. Upon further review, you discover that because of the recent change in waste management practices, workers could be potentially exposed to levels above OELs, and you also find that there are additional workers associated with performing that task that have been exhibiting similar health symptoms. After confirmation of these results, you request that waste management practices be placed on hold until further review is conducted, along with implementing a medical restriction for those workers who appear to be potentially overexposed to toxic contaminants. In addition, you request a consultation with a toxicologist to further understand the impact of exposure to multiple toxic contaminants.

4.6.4 "I Have Been Sick for the Past 6 Months": Lessons Learned

Some of the lessons that can be gleaned from the case study are listed below:

1. At the initial notification by the worker, who suspects his work environment may be causing his illness, you should have completed a thorough investigation that included activities such as:
   a. Being responsive to the questions based on the experience of the worker and your understanding of changing conditions.
   b. Evaluating historical monitoring data, locations of collection, and proximity of work tasks being conducted.
   c. Determining what by-products were generated in the process and conducting a full-shift monitoring campaign to determine if any waste by-products could be harmful to workers.
2. Employees may not believe you had thoroughly investigated the health concerns about their safety.
3. You may not have thoroughly understood the process changes associated with the change in management of waste and the potential to generate an additional chemical hazard.

It is important that industrial hygienists respect and treat the workforce as equals in the workplace because oftentimes the workers are generally the most knowledgeable in the tasks that they perform on a daily basis.

Questions to Ponder for Learning

1. What are the potential company and professional liability costs associated with regulatory noncompliance and realized health impacts?
2. What is the difference between a quantitative and qualitative risk assessment, and how are they used in the reduction of health risks in the workplace?
3. Why is it necessary to implement a robust monitoring and sampling plan? What are the factors that should be considered when developing a plan?
4. What are some typical documents used when conducting a quantitative assessment?
5. How has the relationship between the industrial hygienist and the workers evolved?
6. What additional actions could the industrial hygienist have taken to reduce the overall health risks associated with the work process?
7. What actions can the industrial hygienist take, in the future, to fully evaluate all contaminants to which the worker was exposed?
8. List the reasons why exposure monitoring should be performed.
9. What are the limitations of using vapor monitoring badges?
10. Explain the advantages and disadvantages of handheld electronic monitors.

REFERENCES

1. OECD, 2022. *Technical Report. Establishing Occupational Exposure Limits*, Series on Testing and Assessment, No. 351. Environment Directorate Chemicals and Biotechnology Committee. https://one.oecd.org/document/ENV/CBC/MONO(2022)6/en/pdf#:~:text=The%20Occupational%20Exposure%20Limits%20%28OELs%29%20are%20set%20by,limits%20or%20as%20guidelines%2C%20and%20by%20professional%20organisations
2. Deveau M, Chen CP, Johanson G, Krewski D, Maier A, Niven KJ, Ripple S, Schulte PA, Silk J, Urbanus JH, Zalk DM, Niemeier RW, 2015. The Global Landscape of Occupational Exposure Limits--Implementation of Harmonization Principles to Guide Limit Selection. *Journal of Occupational and Environmental Hygiene*, 12 (Sup 1): S127–44. doi:10.1080/15459624.2015.1060327. Erratum in: J Occup Environ Hyg. 2016 Nov;13(11):D217. PMID: 26099071; PMCID: PMC4654639. https://www.ncbi.nlm.nih.gov/pmc/articles/PMC4654639/
3. NIOSH, 2019. Technical report: The NIOSH occupational exposure banding process for chemical risk management. By Lentz TJ, Seaton M, Rane P, Gilbert SJ, McKernan LT, Whit-taker C. Cincinnati, OH: U.S. Department of Health and Human Services, Centers for Disease Control and Prevention, National Institute for Occupational Safety and Health, DHHS (NIOSH) Publication No. 2019–132, https://doi.org/10.26616/NIOSHPUB2019132.
4. AIHA, 1998. *A Strategy for Assessing and Managing Occupational Exposures*. AIHA Press, ISBN 0-932627-86-2

5 Risk-Based Industrial Hygiene

5.1 INTRODUCTION

Today risk is a topic of great importance across the globe because it often translates into dollars and cents. Many world leaders are attempting to address their level of risk tolerance and seeking ways to minimize the risk that their company is subjected to in the face of being challenged with many life-altering events and increased regulatory oversight and associated fines.

Risk is not an easy topic to discuss or concept to define, since a diversity of factors are involved in defining and quantifying risk; these factors are both situational and people dependent. Defining allowable risk is even more complicated when the health and well-being of people are involved. In addition, risk can take on different meanings depending on the context in which it is being used. For example, financial risk is defined in general terms as the probability that the return on investment will be lower than expected; in the workplace, risk can be defined in terms of the consequence and probability of a hazardous event or activity; and in the food industry, risk is generally defined as the possibility that a certain food hazard will have a negative impact on people or the environment.

As leaders across the globe make an attempt to define the level of risk that they are willing to consider as appropriate for their business and workers, they are also having to take into consideration the many regulatory requirements that have been passed to guide some of these risk levels and decisions. Seeking to define and minimize risks, when it can be a matter of life or death for individuals and workers, is not an easy task. A good example of the power of risk acceptance, outside of the work environment, is driving by a serious car accident. Some drivers immediately stop and will become involved in initial rescue efforts, such as removing a person out of a burning car. Other drivers will not stop but may make notifications to authorities, while others may just stop their car and observe response actions being performed by other drivers. There are different factors that drive risk acceptance, and the field of industrial hygiene is focused on mitigation of both acute and chronic health risks that may occur in the workplace. The industrial hygienist seeks to define and minimize risks, to the lowest level possible, to ensure that people are able to work, without being exposed to conditions that will cause workplace illnesses or injuries. The goal is that every worker return home in the same condition in which they reported to work each day.

DOI: 10.1201/9781032645902-5

5.2 IMPORTANCE OF RISK ASSESSMENTS AND A RISK-BASED APPROACH TO HAZARD MANAGEMENT

The primary goal of evaluating and determining health risks, and ultimately conducting a risk assessment, is to remove hazards and reduce the level of risk by eliminating the risk or adding control measures to help create a safer workplace. Risk assessments are an integral part of a good occupational health and safety management program and are most effective when conducted in the design phase of a project. A robust risk assessment process can

- Create awareness of hazards and the potential associated risks
- Prioritize hazards and the controls needed
- Identify the population that may be at risk
- Determine if existing controls are adequate
- Reduce workplace incidents
- Reduce cost by taking on a proactive approach to incident reduction
- Create awareness and train employees
- Instill trust and credibility in the industrial hygiene and safety and health programs

Companies that have taken and continue to take advantage of ensuring that risk assessments are an integral part of their hazard assessment and control program have seen the value from both monetary and cultural aspects. Risk assessments have proven to be of value from a worker protection perspective; are financially beneficial, such as providing an approach to minimize the company from overinvesting and the costs associated with recovery from incidents; and increase the trust and acceptance of risk by workers.

5.3 IDENTIFYING AND CONTROLLING WORKPLACE RISKS

Conducting risk assessments is the cornerstone of a comprehensive industrial hygiene program. Risk assessments to the industrial hygienist are the equivalent of an engine to a car; without a thorough understanding of risk, working conditions in our society cannot improve. Minimizing workplace risks is a vital part of the ability to control and manage workplace hazards and to protect workers from those hazards.

Corporate leaders, as well as institutions and large and small business owners, must understand that in order to manage the health and safety of workers, health risks must be controlled in every aspect of the business. In order to develop and implement a comprehensive and effective risk strategy, a review and analysis of all work activities conducted within the course of conducting business that can be potentially harmful to workers must occur, including a comprehensive evaluation of the company's internal work culture. Approaching the risk assessment process with comprehensiveness in mind and actions will enhance the effectiveness of the outcome. Important steps to include in this approach are shown in Figure 5.1. These steps are an ongoing part of the risk assessment process.

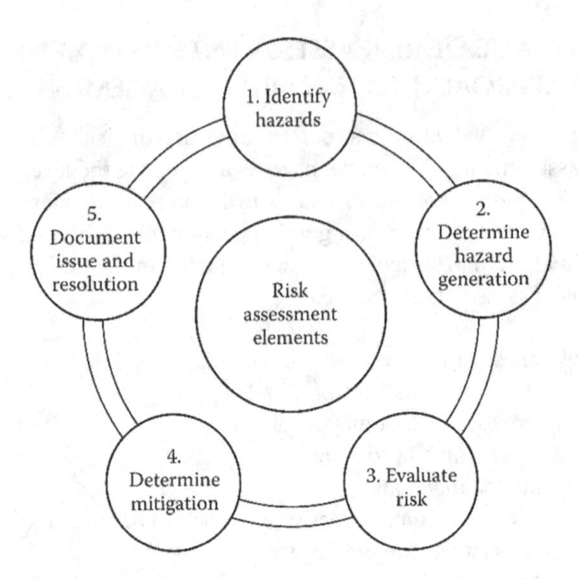

FIGURE 5.1 Basic risk assessment elements.

Comprehensively, the risk assessment process related to the industrial hygiene profession should include:

- Identifying cumulative hazards posed to the worker
- Determining the generation and origination of the hazards
- Evaluating risks posed by the hazards, from both individual risks posed and cumulative risks that can result in an overall higher risk profile
- Identifying and mitigating total risk
- Documenting risk resolution and residual risk to the worker

When identifying whether a correlation exists between an exposure and an associated health risk, one must not lose sight of two important factors that can impact the accuracy of the risk decision: variability and uncertainty.

Variability relates to the range of exposure potential, such as the impact of an exposure from one person to another impacted by factors such as preexisting conditions, genetic factors, or enhanced sensitivities; uncertainty refers to the inability to ensure 100% accuracy, sometimes due to the lack of data or run time. Risk to a worker is generally determined to be acceptable only when the risk is understood, mitigated, or determined to pose little to no potential health hazard.

5.4 ADDRESSING INDUSTRIAL HYGIENE RISKS IN THE WORKPLACE

Assessing industrial hygiene risks in the workplace is multifaceted and takes on many forms. It is not always feasible to determine these risks using one process or system or reviewing one type of documentation. Risk assessments that are explicitly designed to ensure worker safety and health are dynamic and an ongoing process. In addition, these types of risk identification processes integrate concepts in the areas of science, engineering, math, and professional judgment, along with the work experience of the industrial hygienist.

Developing and implementing a risk management program can be especially challenging for the industrial hygienist when there is little or no data available or lack of support from the leadership team. Leadership sets the stage for what is viewed as valuable to the organization. Admittedly, some leaders may at times prefer productivity and may be willing to accept a higher level of risk to achieve these increased productivity levels. At times, productivity may be impacted when there is a high-risk activity being performed primarily because of hazard controls that are in place, or the task may need to be performed at a slower pace to protect workers from injury. Because of the possibility of these challenges, it is important that the industrial hygiene function reports to a senior manager to ensure that management support and resource needs are available as needed.

5.4.1 Industrial Hygiene Risk Assessment

Risk assessments are performed daily by an industrial hygienist to facilitate the safe performance of work. The resulting decisions are generally not determined in a vacuum because of the importance of the decisions, as they are directly connected to protecting the health, safety, and well-being of a worker. When evaluating risk and incorporating risk into the industrial hygiene work process, industrial hygienists will generally conduct risk assessment by two approaches. As depicted in Figure 5.2, these approaches are known as quantitative and qualitative risk assessments. Of the two, quantitative assessments are generally preferred since the decision is made using data that represents the activity, process, or work environment. However,

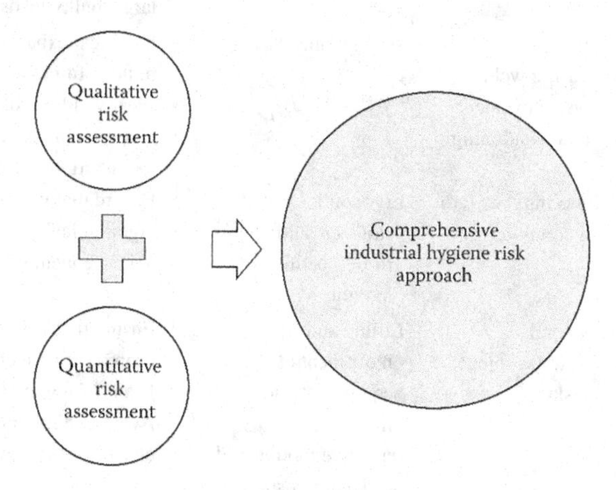

FIGURE 5.2 Industrial hygiene risk assessment approach.

together the two concepts form the fundamental basis for a comprehensive risk assessment approach, which is most protective of human health and life.

As discussed in Chapter 4, both risk assessment components can be conducted simultaneously or in sequential order. When conducting both analyses, do not be surprised if you have to review some of the same types of information or processes to fully complete each of the assessments and ensure that all risks have been ranked, accounted for, and mitigated. In addition, the industrial hygienist can develop a risk table or register to keep track of each risk. An example of a risk register is shown in Table 5.1. Each hazard on the risk registry should be evaluated to determine the risk level. Below are some information sources that can be used to help determine the risk level:

- Safety data sheets (SDSs) for the chemical or product
- Manufacturer documentation
- Regulatory standards and requirements
- Previous injury, accident, and near-miss reports
- Previous assessment results
- Past worker experiences

TABLE 5.1
Example of a Risk Register

Task	Hazard	Activity Risk	Control
Conducting chemical inventory	Transporting chemical from storage to the process area	Lifting chemical containers	• Keep containers tightly sealed • Ensure that chemicals are labeled and identifiable • Transport compatible chemicals together • Use a lifting jack or forklift for large, bulky loads
Welding	Fumes generated during welding; body movement and positioning	Welding metals	• Use local exhaust ventilation to remove fumes and prevent buildup • Wear welder protective equipment • Have a trained worker performing the function of fire watch
Data entry	Entering data into a computer	Ergonomic-related strains resulting from repetitive motions	• Ensure that workstation is ergonomically friendly • Take frequent breaks
Housekeeping	General housekeeping tasks	Lifting and movement of equipment and materials; repetitive motion and awkward posture	• Ensure that workers are trained in proper lifting techniques • Limit lifting to 50 lb • Avoid coming into contact with cleaning chemicals; wear gloves

5.5 RISK RANKING

Once the hazards and subsequent risk have been identified, it is pertinent that each hazard and associated risk be ranked and catalogued. The ranking process will ultimately shed light on whether the hazard can be acceptable, and work can continue without implementing any mitigating controls, or whether the work must be paused or suspended until controls are decided on and implemented.

There are various risk-ranking tools available to assist in completing the process. It is important that the right process be used to achieve a ranking result that is protective of property, worker health, and the environment. Risk rating is dependent on the likelihood of an event occurring, the severity of the event, and the injury potential to workers. Tables 5.2 and 5.3 represent risk quantification tools that can be used in quantifying risks. When quantifying and ranking risk, a simple formula is often used. Below is an example of a simple risk equation:

$$\text{Risk (R)} = \text{Likelihood (L)} \times \text{Severity (S)} \; (R = L \times S)$$

TABLE 5.2
Example 1 of a Risk Quantification Tool

Likelihood	Severity	OSHA
Highly unlikely	Minimal injury	First aid
Unlikely	Slight injury	First aid
Likely	Serious injury	Recordable
Highly likely	Major or catastrophic injury	Recordable

TABLE 5.3
Example 2 of a Risk Quantification Tool

Risk Band	Control
Minimal (1–3)	Review existing measures (may be able to maintain current measures). Note: Even if the risk is low, every attempt should be made to lower the risk because the health and safety of the worker are at stake
Low (4–6)	Review control measures (may need to tweak or add additional controls that are minimal)
Medium (7–9)	Review and improve control measures
High (10–12)	Review and improve control measures. Discontinue work

5.6 INTEGRATION OF A RISK-BASED CONSENSUS STANDARD INTO INDUSTRIAL HYGIENE

One of the most recognized consensus standards for safety and health is the International Organization for Standardization (ISO) 450001, *Occupational Health and Safety Management Systems.*[1] The ISO standard recognizes safety and health (industrial hygiene) as a distinct and separate management system. The primary reason a company adopts ISO 45001 is because they want to demonstrate they meet an international standard for performance that is risk-based. A company that is ISO 45001 conforming is recognized as meeting defined performance standards, operationally consistent, efficient, and internationally recognized as excellent.

Depending upon how the company is organized, industrial hygiene can be managed as its own management system, or within the safety and health management system as a subprogram. A typical company may implement their industrial hygiene program through a compliance-based approach. For example, a company may only consider OSHA permissible exposure limits in implementing their industrial hygiene program; they do not consider American Conference of Governmental Industrial Hygienists (ACGIH) threshold limit values (TLV) to further minimize occupational exposure of their workers to hazardous chemicals. Often a company may incorporate ACGIH TLVs, but they are not legally required to adopt the more conservative occupational exposure limits.

A company that conforms to the ISO 45001 must identify and mitigate risks to all program elements, whether that is applied at safety and health or industrial hygiene program level. For example, Section 6.1.2.2, *Assessment of OH&S Risks and Other Risks to the OH&S Management System,* requires a company that has adopted ISO 45001 to:

- Assess OH&S risks from identified hazards, while taking into account the effectiveness of existing controls, and
- Determine and assess the other risks related to the establishment, implementation, operation, and maintenance of the OH&S (or IH) management system.

Further, Section 6.1.2.2 states:

> The organization's methodology and criteria for the assessment of OH&S risks shall be defined with respect to their scope, nature, and timing to ensure they are proactive rather than reactive and are used in a systematic way. Documented information shall be maintained and retained on the methodologies and criteria.

As depicted in Figure 5.3, all ISO standards, including ISO 45001, are based on a continuous improvement model: Plan-Do-Check-Act (PDCA). The PDCA model was developed to demonstrate continuous improvement and includes both compliance (e.g., meeting regulatory established requirements) and performance-based criteria (e.g., metrics and non-conformances identified and tracked).

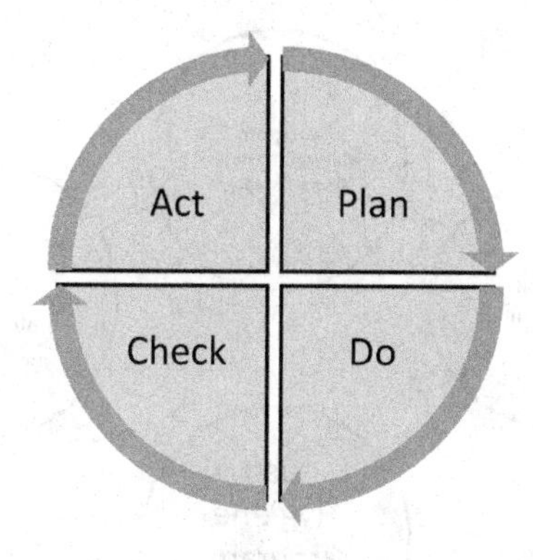

FIGURE 5.3 Plan-Do-Check-Act model.

The industrial hygiene program can be considered a management system and is, in terms of priority, as other company or institution management systems. The PDCA model can be applied to implementing any program or discipline and can be applied no matter the size of the task or program.

As industrial hygiene is safety and health based, integration of ISO 45001 into the industrial hygiene program elements is a natural progression to elevate the overall program to the next level. The applicability of the ISO sections is dependent upon the level of health and financial risk the company or institution desires to mitigate. For example, when implementing an industrial hygiene program at a construction site, the depth to which the ISO 45001 requirements are integrated into program elements would be less than the depth to which ISO 45001 requirements would be implemented at an industrial hygiene program affiliated with a laboratory where toxic biological agents are managed. Development of a crosswalk of the ISO standard against the existing industrial hygiene program can provide insight as to the depth the company desires to integrate ISO into their program.

As discussed in Chapter 1, Figure 5.4 depicts the example of industrial hygiene program elements. An approach for integration of the PDCA model into the industrial hygiene program elements is discussed below. The subsections below were written to provide an example of how the ISO standard could be integrated into the industrial hygiene program.

5.6.1 Industrial Hygiene Program Management and Administration

Specific sections of ISO 45001 requirements which may apply to the industrial hygiene program element Management and Administration include (examples) Section 3, *Terms and Definitions*; Section 4, *Context of the Organization*; Section 5, *Leadership and Worker Participation*; Section 7, *Support*; Section 9, *Performance Evaluation*; and Section 10, *Improvement*. There is a clear relationship in planning of the work or the organization and planning for program management, administration, evaluating program performance, and effective resolution of industrial hygiene issues for continuous improvement. Included within this functional element are items such

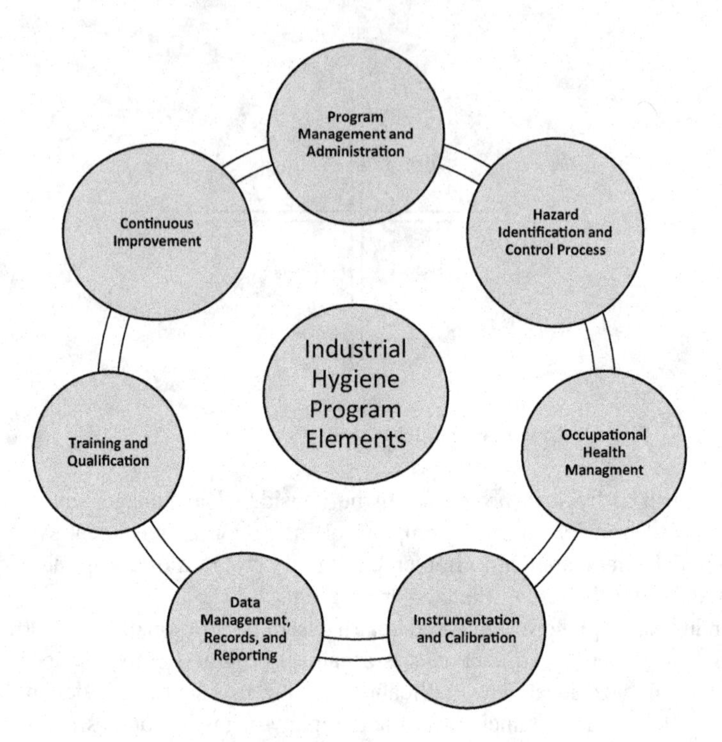

FIGURE 5.4 Industrial hygiene program elements.

as the planning and documenting of leadership commitments, industrial hygiene policies (e.g., ALARA), roles and responsibilities, policies and procedures for worker involvement, safety and health committees, planning for emergency events, and development of metrics useful in identifying negative performance trends.

5.6.2 Hazard Identification and Control Process

The hazard identification and control process is one of the primary processes the industrial hygienist uses to mitigate workplace hazards. The industrial hygienist is actively involved in the identification of general hazards, but also in the identification of job-specific hazards. Applicable sections of ISO 45001 which may be relevant to this industrial hygiene program element include (but not limited to) Section 5, *Leadership and Worker Participation*; Section 6, *Planning*; and Section 8, *Operation*.

The industrial hygienist is intimately involved in the identification and implementation of hazard controls whether the hazard affects the entire company (e.g., COVID-19, air quality), or a smaller work crew (e.g., operational spill). Risks from hazards which are identified, per Section 6 of the standard, are required to be documented and take into account effectiveness of existing controls, and to determine and assess the other risks related to the establishment, implementation, operation, and maintenance of the industrial hygiene management system. The traditional hierarchy of hazard controls is applied, but also the industrial hygiene program implements a process for implementation and control of planned temporary and permanent changes that could impact an industrial hygiene program.

The ISO standard also refers to establishment of a *Management of Change* process (Section 8.1.3) similar to the *Management of Change* process required per 29 CFR 1910.119, *Process Safety Management of Highly Hazardous Chemicals*. Under 29 CFR 1910.119(l), *Management of Change*, companies are required to establish and implement written procedures to manage changes (except for "replacements in kind") to process chemicals, technology, equipment, and procedures, and changes to facilities that affect a covered process. If the industrial hygienist works in the Unites States and performs work at a facility that manages highly hazardous chemicals, they can use the *Management of Change* process established to meet the federal regulation within their ISO 45001 Industrial Hygiene Program.

5.6.3 Occupational Health Management

This industrial hygiene program element refers to management of medical and surveillance programs, company health wellness programs, and management of workers on medical work restrictions; activities the industrial hygienist may be expected to be involved in program implementation. A company's Occupational Medical Health Program is typically managed as its own program element and application of the ISO 45001 requirements is tailored to the level of risk (often financial) owned by the company or institution. ISO 45001 requirements typically relevant to the occupational health program include Section 4, *Context of the Organization*; Section 5, *Leadership and Worker Participation*; Section 6, *Planning*; Section 8, *Operation*; Section 9, *Performance Evaluation*; and Section 10, *Improvement*. The industrial hygienist is involved with all the sections listed above.

5.6.4 Instrumentation and Calibration

Application of ISO 45001 for the industrial hygiene program element starts with the identification of the quality standards applicable to industrial hygiene instrumentation, calibration, and laboratory. Sections of ISO 45001 which could be relevant to instrumentation and calibration include Section 6, *Planning* and Section 8, *Operation*. The identification of the quality standards applicable to the instrumentation program should be clearly defined and then procedures developed to demonstrate implementation. Risks to the instrumentation and calibration program should be identified and mitigated (e.g., instrument interferences).

5.6.5 Data Management, Records, and Reporting

Included within this industrial hygiene program is the data generated from industrial hygiene monitoring, any type of documentation or technical basis needed to implement a program or activity, and reporting of data to both the employee and for the company. ISO 45001 sections potentially applicable to this program element include Section 7, *Reporting*; Section 8, *Operation*; and Section 9, *Performance Evaluation* (may be applicable depending upon how organized in the overall program). Example work activities where Section 7.5.3, *Control of Documented Information,* could apply would be the management of field monitoring records and their distribution, access, retrieval, and use.

5.6.6 Training and Qualification

Applicable sections of ISO 45001 relevant to the industrial hygiene program element, *Training and Qualification,* may include Section 4, *Context of the Organization*; Section 8, *Leadership and Worker Participation*; and Section 8, *Operation.* Risks when executing the company mission using unqualified workers and unqualified industrial hygiene personnel should be identified, understood, and managed. Using inadequately trained personnel to implement the industrial hygiene program often leads to an inadequate technical basis for industrial hygiene decisions, as well as can lead to significant non-compliances to regulatory requirements and an increased risk posture for the company.

5.6.7 Continuous Improvement

The very basis for adopting a consensus standard, such as ISO 45001, is to exhibit continuous improvement of the safety and health or industrial hygiene program. Specific sections of ISO 45001 related to continuous improvement of the industrial hygiene program may include Section 4, *Context of the Organization*; Section 9, *Performance Evaluation*; and Section 10, *Improvement.* In particular, Section 10 of the standard specifically focuses on the identification of opportunities for improvement and nonconformances, including the non-fulfillment of a requirement. For example, with respect to industrial hygiene, a nonconformance can be caused by not meeting a PEL for chromium. The industrial hygiene organization would have a defined process for documenting the nonconformance, evaluation of a nonconformity, and the identification and implementation of corrective actions which would resolve the nonconformant condition. Also included within Section 10 is the continual improvement of the culture that supports the safety and health or industrial hygiene program.

5.7 RISK COMMUNICATION

Communicating risks, in particular health risks posed from the performance of work activities and processes, are just as important as defining and mitigating the impact of the risk. In addition to having basic knowledge of and understanding the impact of the risk, the industrial hygienist should couple the information with good communication skills so that the communicator can gain acceptance of the risk from and the respect of the workforce. Therefore, the trust and credibility of the person delivering

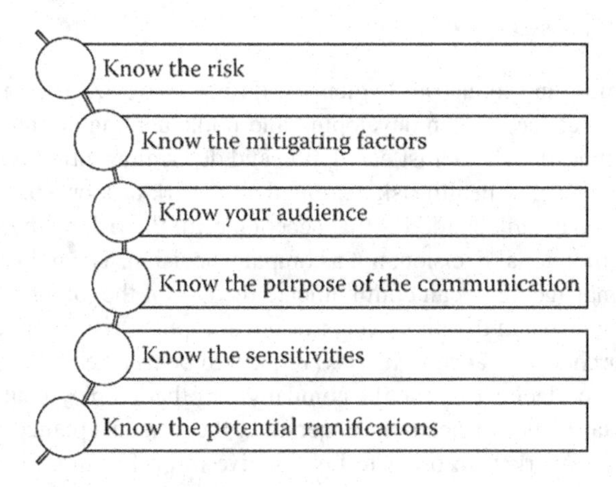

FIGURE 5.5 Risk communication principles.

the message are paramount. If the person responsible for delivering the message is not viewed as trustworthy and sincere, the message will be questioned and may even fall on deaf ears.

Risk communication must be a balanced approach. The principles listed in Figure 5.5 are important to developing and delivering the risk message and gaining acceptance for the message.

1. Know the risk: Knowing the risk provides insight into key attributes and the appropriate means needed to deliver a message that can be believed and trusted by the listeners.
2. Know the mitigating factors: Mitigation factors are actions taken to remove or control the risk to prevent harm to people, the environment, or equipment. Knowledge of mitigation factors can aid in communicating that the risk is under control and not expected to produce harm to the workers.
3. Know the audience: Knowledge of the audience provides insight into the way the communication should be constructed and delivered.
4. Know the purpose of the communication: The purpose of the communication is instrumental in shaping the tone and content of the message to be delivered. The tone and content of a message ultimately determine how it will be received by others.
5. Know the sensitivities: Often, there are some workers who are sensitive to a particular issue or circumstance. When communicating risk or issues that can create stress, care must be given to the delivery.
6. Know the political ramifications: Most communication related to risk and human health has some type of political ramifications, which can include loss of trust by workers and the public and potentially the fear of continuing to work for the company.

5.8 RISK ACCEPTANCE

Often, companies, and industrial hygienists, do not recognize that risk acceptance is a fundamental concept when developing and implementing an industrial hygiene program. The industrial hygienist can define and determine, either quantitatively or qualitatively, the level of health risk exhibited by a work task or work process; however, what is often misunderstood is that acceptance of the risk by the worker is both a personal and individual decision, not a company decision. The industrial hygienist, and company management, can communicate the risk of the job or work activity to workers, but the decision to either accept or not accept the risk is up to the workers; they will either choose to accept the risk, request to be transferred to a different job in the company, or decide to leave the company for other employment.

The industrial hygienist needs to understand the risk acceptance process of the company, and the worker, in order to be effective in performing his or her profession. Worker engagement and partnership are among the most effective and powerful methods in achieving acceptance of risk by the worker; however, in order for the method to be effective, the company must recognize, acknowledge, and actively conduct business in a manner that is reflective of the worker being the final decision maker in the risk acceptance process. Acceptance of risk, and in particular acceptance of increased risk, is an iterative process that must be managed in a respectful manner, which does not happen overnight.

The culture of the work environment is highly influential in the acceptance of risk by the worker, and workers are intuitively aware of managers and supervisors who are not genuine and do not display personal character attributes such as integrity and trust. In particular, it is of great importance that managers and supervisors understand that they may be in charge of managing work activities, but it is the individual worker who has a personal responsibility to ensure that his or her safety and well-being are not being compromised in order for the company to profit. If properly trained, the industrial hygienist can be a conduit to drive and achieve greater risk acceptance and improve the overall culture of the work environment.

QUESTIONS TO PONDER FOR LEARNING

1. What role does the industrial hygienist play in the risk assessment and mitigation process?
2. How should an industrial hygienist approach the risk assessment process?
3. What are the elements of the risk assessment process?
4. What two factors can impact the accuracy of a risk decision?
5. Name the information sources that can be used to aid in the determination of risk levels.
6. Explain the importance of peer reviews when making risk-based decisions.
7. What role does trust play in making risk-based decisions?
8. What is the purpose of adopting a consensus standard for an industrial hygiene program?

9. Identify and discuss integration of a consensus standard into an industrial hygiene program.
10. Identify and discuss methods to achieve risk acceptance by the worker.

REFERENCE

1. International Standard, ISO:45001, 2018. *Occupational Health and Safety Management Systems – Requirements with Guidance for Use*, International Organization for Standardization.

6 Recognizing, Evaluating, and Controlling Workplace Hazards

6.1 INTRODUCTION

Health and safety hazards exist in every workplace. History has shown that workers are injured in every workplace, regardless of the business type, and work activities being performed. Some hazards in the workplace are easily identified and corrected, while others are difficult to identify and control, creating extremely dangerous conditions that could be a threat to life or long-term health. The best way to protect against the hazards that are inherently present in the workplace is to know how to recognize and control them to protect workers from being negatively impacted in the workplace.

The industrial hygienist is one of the primary professionals involved with the workforce in the identification and recognition of hazards in the workplace. Traditional industrial hygiene training focuses primarily on chemical and biological hazards; however, depending on the size of the company or institution by which the industrial hygienist is employed, the recognition of workplace hazards may go beyond the traditional industrial hygiene field because the job may also include functioning as a safety manager or engineer, environmental manager, or operations specialist. Recognizing or identifying hazards is one of three steps used to protect the worker from hazards in the workplace. Along with identifying the hazards, the industrial hygienist must analyze them to determine the health risk to an employee, along with understanding how to control the hazards.

Some hazards may exhibit more than one risk factor. For example, exposure to noise may not only cause hearing damage, but also interrupt or completely impact and stop communication between workers, and may be distracting, which can lead to an accident and potentially an injury. Another example to consider is the use of metal uranium; not only is it linked to cancer, but also the chemical is a health risk to the kidneys. Without understanding the health or workplace risks associated with the work to be performed, methods used to analyze and control the hazard, and essentially risk to the workers, may not be effective. Figure 6.1 shows the relationship between the hazard identification process and the overall work control process.

Physical hazards are the most commonly encountered hazards in the workplace and refer to material or equipment that, if left unmitigated, has the potential to harm a person depending on the work task or environmental conditions. Common examples

DOI: 10.1201/9781032645902-6

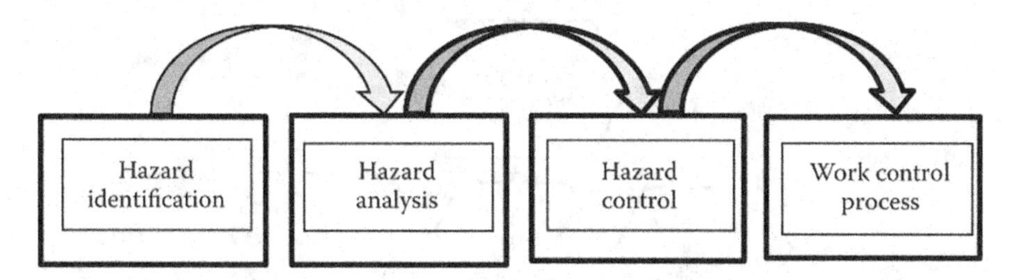

FIGURE 6.1 Hazard identification and work control process relationship.

include slick floors, uneven walking surfaces, a fall from heights, motorized vehicle accidents, and exposure to a blade or sharp object (unguarded). Physical hazards also include energized material and equipment.

Chemical hazards refer to the ability of chemicals or agents to cause physical harm to people, such as burns or cancer. Chemical hazards may include organic, inorganic, or metal compounds, and harm may be caused by either dermal contact or ingesting or inhaling the chemical. Common examples of chemical hazards in the workplace include asbestos, lead, beryllium, ammonia, nitrates, and carbon tetrachloride.

Biological hazards refer to the ability of biological pathogens to be transmitted through dermal contact or inhalation. Common examples of biological hazards include viruses, bacteria, fungi, blood, and human feces. Biological hazards are gaining increasing importance in today's workplace because of a heightened awareness among the general public and employees who have suffered from the long-term health risks posed by biological hazards, such as the Coronavirus-19 (COVID-19).

Radiological hazards may include both ionizing and nonionizing radiation, which can be harmful to humans through general exposure, inhalation, and ingestion. Common examples of radiological hazards include microwaves, diagnostic x-rays, and work processes that use or generate radiological contaminants or non-destructive magnetic examination techniques that use radiation.

An effective hazard identification and control process is based on thoroughly understanding the inner relationship between the three process elements of the hazard identification, analysis, and control process. In conjunction with the worker, the industrial hygienist will identify the hazard and then determine methods by which to analyze and control the hazard. When the hazard is controlled and/or removed, the industrial hygienist must understand how the risk posed by the hazard was reduced and what level of risk remains to the exposed worker from other hazards. At each step of the process, information is gathered and continuously used to improve controlling the hazard and reduce the level of risk posed to the worker. In order for the process to be effective, the workers must also understand how they are being protected, and the reduction of health risk should be effectively communicated. Since it is recognized that hazards exist in every work location, below are some tools the industrial hygienist can use to improve the hazard recognition process. Figure 6.2 depicts a summary of common tools used for recognizing workplace hazards.

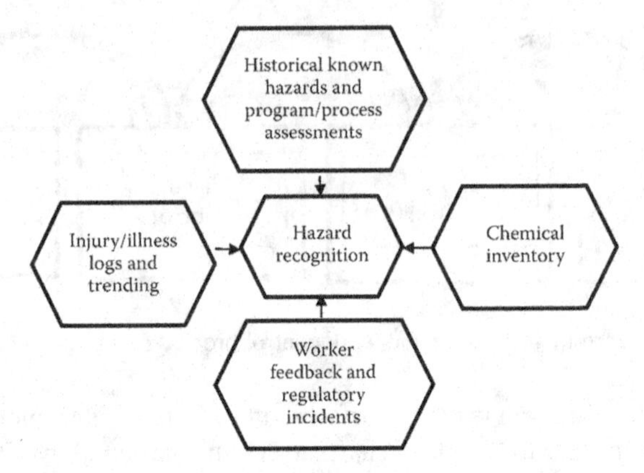

FIGURE 6.2 Hazard identification tools.

Depending on the work function performed, workplace hazards may include hazards that are stationary in the work environment or are introduced by the work being performed. The practice of hazard recognition is a fundamental work process that must be performed on a daily basis, and depending on the work function performed, the workplace hazards will vary. An underlying premise for all the hazard recognition mechanisms is worker feedback.

Employers have the responsibility to ensure that workers are protected against workplace hazards. In addition, workers have the right to be informed of potential hazards in the workplace and can refuse to perform work as a result of the hazards involved. If an employee refuses to perform a task because performing the work is believed to be hazardous, employers have no legal recourse and are not legally supported if retaliation occurs. Employers that provide safe working conditions for their employees are on the road to having a successful and prosperous business; workers are very supportive of employers, and become loyal, when the company is concerned about their well-being.

Many employers provide hazard recognition training to workers at various stages of their employment. Hazard recognition training should be considered when:

- An employee is onboarded with the company. New employees are generally receptive and eager to learn about the company and its value, as well as the workplace culture.
- An employee is reassigned to a new job within the company.
- An employee is asked to complete a new task.
- An employee is asked to perform a task that is of high risk and high consequence that has not been performed in recent times.

Recognizing that not all hazards in the workplace will cause injuries, it is incumbent upon the employer to make every effort to mitigate hazards and ensure that workers are protected when performing work. An integral part of ensuring worker safety is to have a process in place that is effective in identifying and mitigating hazards before work begins. The hazard identification and mitigation processes are not a one-person task or responsibility. It requires support and input from supervision, workers, and subject matter experts.

6.2 HISTORICAL CHEMICAL AND INDUSTRIAL HAZARDS

There are many types of hazards that can be found in the workplace. The types of hazards are based primarily on the type of business the company is engaged in; however, there are some common hazards that are germane to every workplace. An example of some common workplace hazards is listed in Table 6.1.

In addition, hazards can be generated from poor work practices and other human behaviors. Poor work practices, such as not following procedures, or not watching for others when performing work, can create workplace hazards that can prove to be life changing for workers. These life-changing events can cause injuries and illness in the workplace that may be difficult or, in some cases, impossible to recover from. Examples of poor work practices and human behaviors that can contribute to unsafe work practices are:

- Using unsafe or defective tools
- Using tools or equipment in ways for which they were not intended
- Operating equipment or vehicles at unsafe speeds
- Using improper lifting techniques

TABLE 6.1
Common Workplace Hazards

Hazard	Description
Chemical	Exposures to chemicals, such as cleaning products and solvents, while performing preparation tasks, such as chemical mixing or fueling stations
Physical	Tripping hazards, excessive noise, damaged electrical cords, unguarded machinery and rotating devices, vibrations, and working at heights (ladders and scaffolding)
Biological	Exposure potential arises when working with infectious people, animals, or plants; typical exposures are from blood and other body fluids, bacteria or viruses, insect bites, and animal or bird droppings
Ergonomic	Occur when the task being performed facilitates placing the body into a position of strain; examples include poor lighting, improperly designed or adjusted workstations, repetitive movement, and frequent lifting
General safety	Some of these hazards result from poor housekeeping, while others can result from workers tripping while walking with no apparent hazardous conditions in play
Psychological	Can result from personal or workplace stress

- Removing or disabling a protective guard or safety device from equipment or tools
- Failing to wear protective equipment
- Standing or working under suspended loads
- Working on or utilizing equipment for which training has not been completed
- Poor housekeeping
- Performing tasks that have not been evaluated and released to be worked (scope creep)
- Inadequate bounded work scopes
- Lack of worker focus on tasks and the associated hazards
- A renegade attitude by workers and supervision

The health effects of exposures to chemicals and toxic materials in the workplace, and the associated physiological and emotional impacts, have been well documented over the years for many industries. Since the beginning of the industrial age, the hazards associated with particular industries or professions have been recognized and mitigated. In particular, health risks associated with direct exposure to chemicals have become an increasingly recognized workplace hazard in today's work environment. No matter what the industry or business, chemicals are used every day, and while workers understand that they are required to handle chemicals in performing their job, health risks posed by these chemicals are becoming less accepted, by the workforce and the public, as they become more aware of how these hazardous chemicals can impact their health and possibly the health of their families.

If the company performs work in an older building, or works with asbestos or lead, then the industrial hygienist should be aware of how these, and other historically hazardous contaminants, interact with the human body. For example, friable, crushed asbestos has been known to cause a lung disease, asbestosis, when inhaled; fibers settle in the lower part of the lung and cause scarring, making it difficult for people to breathe and obtain needed oxygen. Lead is another example of a contaminant that is historically known to cause serious health impacts. Lead affects many organs and body functions to some degree when inhaled or ingested; however, it is most recognized to be a neurological and bone hazard.

Through the use of the internet, today's workforce is much more aware of the health risks posed by chemicals that are used by them every day in performing their job. One of the best resources for workers and the public to reference in understanding how chemicals can impact them, within their own residential area, is the U.S. Chemical Safety and Hazard Investigation Board (CSB), website: http://www.csb.gov.[1]

The CSB is an independent federal agency empowered to investigate chemical accidents that occur across the United States. The focus of the CSB is on industrial accidents, ranging from petroleum accidents, such as the Deepwater Horizon accident that killed 11 people, to explosions in fertilizer plants across the United States. The CSB does not issue fines, but rather works with other regulatory agencies, such as the Occupational Safety and Health Administration (OSHA) and the

Environmental Protection Agency (EPA), in understanding what caused an industrial accident and how it can be prevented in the future. In 2002, the CSB's first hazard investigation on reactive chemicals included a review of more than 150 serious accidents involving uncontrolled chemical reactions in industry. This investigation led to new recommendations to OSHA and EPA for regulatory changes. A second hazard investigation on combustible dusts is currently in progress that is expected to impact how OSHA regulates industries that generate and mitigate combustible dusts and provides better hazard control for worker.

The challenge for the modern industrial hygienist is to recognize that workers are more educated and aware of the hazards, and the industrial hygienist must work with the workforce in collectively identifying and managing the hazards. For example, ammonia has long been recognized as a hazardous chemical, but it is easily recognized by smell, before its hazardous properties can cause harm to the human body. Although workers may accept the smell of the chemical at lower concentrations, they may become concerned that over time breathing that concentration may cause chronic (long-term) health effects. Also, some workers may be sensitive to the chemical and may exhibit acute (short-term) health effects that the industrial hygienist will have to address.

The industrial hygienist is challenged with adequately communicating the recognition of the chemical hazard, along with communicating how the hazard will be adequately mitigated and the health risks to the worker minimized. In addition to the hazards of asbestos and lead, other common occupational illnesses associated with chemicals, and other physical and biological hazards, are listed in Table 6.2. The table represents a short list of well-known workplace illnesses; however, there are many more that are present in the workplace depending on the chemicals and physiological and biological contaminants being used in the manufacturing or industrial work process.

TABLE 6.2
Example Workplace Illnesses

Contaminant	Symptoms and Illness
Beryllium	Skin sensitization; berylliosis; health impacts to the liver, kidneys, and heart; neurological disorders; lymphatic system problems
Silica	Lung diseases, such as bronchitis, emphysema, and chronic obstructive pulmonary disease; kidney and immune system diseases
Anhydrous ammonia	Irritation to skin and eyes, shortness of breath, headaches, bronchitis, seizures
Carbon tetrachloride	Headaches, dizziness, liver and kidney damage, neurological disorders
Benzene	Headaches, dizziness, skin irritation, liver, and kidney damage
Acetone	Headaches, dizziness, disorders with cognitive thinking, neurological disorders
Chlorinated hydrocarbons	Headaches, dizziness, disorders with cognitive thinking, neurological disorders, liver, and kidney damage

The industrial hygienist should be intimately familiar with the work equipment, processes, and chemicals that are introduced and used in the work environment. Therefore, the industrial hygienist must understand what chemicals and toxic materials are being used, in what form they are being used, and any historical incidents that have occurred in the work environment. Feedback from workers, engineering design information, along with historical information regarding toxicological information in the general industry, is useful in determining whether the hazard posed from the chemical or toxic material is understood and recognized not just by management, but also by the workers. In many industries, the chemicals and toxic materials used in a process have been identified and are in some form being managed. Additionally, there are known industrial hazards that have been established based on the industrial process. For example, not only is the industrial hygienist concerned with the toxicological effects from a particular chemical used, but also other hazards, such as physical and biological hazards, must be recognized, and the risk posed by these hazards must be factored into overall worker protection and controls.

6.3 CHEMICAL, PHYSICAL, BIOLOGICAL, AND INDUSTRIAL HAZARDS OF THE PAST DECADE

Over the past decade, industrial accidents have continued to occur and impact local and regional communities. Sometimes the accidents are naturally made, such as the impact of hurricanes, but most often the accidents have been due to human error or faulty equipment (often due to running equipment until failure). As equipment ages, so too does the effectiveness of engineered controls. Another problem that is often seen is that management of a company may choose to defer preventative or corrective maintenance on equipment that, when fails, can cause catastrophic consequences or the impact of the failure is not fully understood. The accidental release of chemicals is often chronicled in the news or on the internet. A recent incident in the United States, whose impact is still being understood, was the train derailment in East Palestine, Ohio.

On the evening of February 3, 2023, a Norfolk Southern train was traveling east carrying 38 cars when the train derailed. Included within the train inventory were 11 cars which contained hazardous materials such as vinyl chloride, butyl acrylate, ethylhexyl acrylate, and ethylene glycol monobutyl.[2] Some cars spilled their loads into an adjacent ditch that feeds Sulphur Run, a stream that joins Leslie Run, which eventually empties into the Ohio River. As a result of the derailment, a fire ensued which further damaged the train cars. First responders implemented a one-mile evacuation zone that affected 2,000 residents. Within two days responders had mitigated the fire, but one of five derailed tank cars, carrying vinyl chloride, started increasing in internal temperature, which was believed to be indicative of the tank undergoing a polymerization reaction, which could lead to an explosion and further spread the harmful fumes. Over the next few hours responders performed a controlled venting, of the five vinyl chloride tank cars, and expanded the evacuation zone up to two miles and dug ditches to contain the vinyl chloride liquid. Although the accident is still under investigation to determine the cause of the derailment, the long-term impact of slowly releasing vinyl chloride into the community, and surrounding areas, are still being evaluated; however, many people within the East Palestine community continue to feel adverse health effects and struggle to work and provide for their families.

One of the more recent significant work-related releases of hazardous chemicals in the United States was the Kingston Tennessee Valley Authority (TVA) Coal Ash Spill.[3] On December 22, 2008, a failure occurred on the northwest side of a dike used to contain coal ash. Subsequent to the dike failure, approximately 5.4 million cubic yards (CYs) of coal ash was released into Swan Pond Embayment and three adjacent sloughs, eventually spilling into the main Emory River channel. A byproduct of coal burning power plants, fly ash contains a variety of metals and other elements which, at sufficient concentrations and in specific forms, can be toxic to biological systems.

On May 11, 2009, TVA entered into an Administrative Order on Consent (AOC) with the EPA Region 4 Office, under the regulatory authority of the Comprehensive Environmental Response, Compensation and Liability Act (CERCLA), to address the coal ash released to the environment.[4] Cleanup of the spill continued over the next few years and was deemed complete in 2015; however, several employees of the engineering firm hired to cleanup the spill developed illnesses, including several cancers and leukemia. In the 10 years since the spill occurred more than 30 workers have died. In 2018 a federal jury determined the engineering firm hired to perform the cleanup had failed to "exercise reasonable care" in protecting workers and likely caused the poisoning of the workers. As a result of being found responsible for the poisoning, workers are able to receive money to cover their damages, including medical testing for all laborers who worked at the site and medical treatment for themselves and their family. One of the more interesting facts of the Kingston Coal Ash event was that over time levels of radiological constituents in the river ash and sediment were observed to be slightly higher than those in ash taken from the Dredge Cell and embayment. Levels of Cesium-137 were much higher in the river ash and sediment. Cesium-137 is a legacy constituent from historical releases from the Department of Energy in the Clinch River. For those workers who directly worked within the river sediments performing cleanup activities, they are managing not only exposure to coal ash, but also potentially exposed to low levels of radiation.

A recent analysis of data collected by the Environmental Protection Agency (EPA) and non-profit groups has determined that accidental releases, through train derailments, truck crashes, pipeline ruptures, or industrial plant leaks and spills, are happening consistently across the country. By one estimate the incidents are occurring, on average, every two days.[5] In the first seven weeks of 2023 alone, there were more than 30 incidents recorded by the Coalition to Prevent Chemical Disasters,[6] roughly one every day and a half. In 2022, the coalition recorded 188, up from 177 in 2021. The group has tallied more than 470 incidents since it started counting in April 2020. Given the prevalence of accidents and releases of hazardous materials, the industrial hygienist needs to have a thorough understanding of toxicity and health effects of all chemicals used in the workplace. In addition, the industrial hygienist must be diligent in identifying all hazards which could work together in creating an even greater hazard.

6.4 WORKPLACE HAZARD INVENTORIES

In the quest to better protect workers, it is a good idea for an industrial hygienist to complete and make available to every worker a facility and health hazard inventory. These inventories should contain pertinent information on the types of processes, along with the known hazards, and the control methods that were used to mitigate or provide the means for workers to protect themselves from being injured. Generation of a facility and health hazard inventory is a good tool for the industrial hygienist because not all information associated with hazards recognition is contained within one inventory or program location. These inventories can be key in planning work in particular for routine work, determining priorities for monitoring and process change evaluations.

6.4.1 Task Hazard Inventory

A comprehensive task hazard inventory is a great tool to provide needed information on hazards and protective measures that can be used in planning and implementing work safely in the pursuit of reducing health risk. A comprehensive inventory contains all known tasks and is updated each time a new task is developed. However, used alone this type of inventory does not provide a complete picture of risk and the types of potential hazards faced by workers. An example of a task hazard inventory is provided in Table 6.3.

TABLE 6.3
Task Hazard Inventory

Task	Hazard	Control	PPE
Welding	Fire	Fire extinguisher training for workers; f ire extinguisher inspected and available; f ire warden available while welding	Welder PPE: include welder shield, gloves, and apron; fire warden PPE: Include safety glasses, hearing protection (if needed), safe distance away from work, and welder screen
Lifting containers and supplies	Sprains and strains	Training (ergonomic, proper lifting techniques)	Gloves and back brace
Chemical addition	Exposure to chemicals	Local ventilation; personal monitoring	Chemical gloves, goggles (face shield), and apron

TABLE 6.4
Example Facility Hazard Inventory

Facility	Location	Hazard	Control	PPE
Chemical processing	Second floor chemical addition station area access	Access stairway steep and slick after cleaning	Stairway posted with a sign stating the condition	Wear rubber sole shoes or safety shoes; h old stairway railing when ascending and descending
Chemical processing	Tank receiving	Skin contact and inhalation of fumes	Local exhaust ventilation	Face shields, goggles, apron, and gloves; respiratory protection if required
All areas	All areas	Walking on uneven surfaces	Brief workers; pay attention and wear appropriate shoes	None

6.4.2 Facility Hazard Inventory

A facility hazard inventory (or list) consists of the physical, chemical, biological, and ergonomic factors that are inherent in the work environment that can cause injury or illness to a worker when in that environment performing work. These hazards are a constant in the work environment and therefore must be taken into consideration each time one enters that work area. These hazards must be factored into the work planning and control process when planning how to perform work. Table 6.4 shows a typical facility hazard inventory and can be very useful in documenting non-job specific hazards and associated health risks. This inventory is typically in the format shown or in a database format.

6.5 INJURY AND ILLNESS LOGS AND INSPECTION TRENDING

Injury and illness logs and information generated from routine inspections can be used by the industrial hygienist for researching and understanding hazards that may be associated with the workplace. Within the United States companies are required to report workplace injuries or illnesses that occurred as a result of work being performed at their job sites. The reporting must be on an annual basis and posted in the workplace. Companies performing work in the United Kingdom are required to report injuries, illnesses, and dangerous occurrences to the Reporting of Injuries, Diseases, and Dangerous Occurred Regulations (RIDDOR).[7] In addition, companies performing work in countries who are members of the European Union are required to report occupational accidents and diseases to Eurostat accidents at work, occupational diseases, and other work-related health problems and illnesses. Regulation 1338/2008/EC states that Member States are obliged to supply statistics to Eurostat on accidents at work, occupational diseases, and other work-related health problems and illnesses.[8]

The industrial hygienist may or may not be involved in managing the injury and illness reporting process, but either way, the injury and illness logs can serve as a good resource in understanding the company's highest-risk work. For example, a review of the workplace logs may indicate that over the past eight years, two workers have been diagnosed with hearing loss. Research into the two individuals' working conditions shows that both at one time worked on the same assembly line. Although sound-level measurements indicate that noise levels were below regulatory exposure limits, there is some equipment in the work location that at short intervals generates noise levels above the regulatory requirements. Further investigation by the industrial hygienist may be needed to ensure that workers are not performing tasks that would expose them to the elevated noise levels for extended periods of time.

Another example is that three workers have been diagnosed with skin dermatitis and skin sensitivity, and all three workers perform jobs in the same work location. The industrial hygienist should investigate the work location and determine whether the workers performed the same task, were located for a period of time within the same work location or used a common work location or change area.

In both examples, as with every workplace incident or injury, it is critical to receive feedback from the workers as to whether they all share something in common or whether they can provide additional information that will assist the industrial hygienist in recognizing an unknown hazard. Whether the industrial hygienist researches the formal incident and illness logs or informal documentation of each incident, information can generally be gathered that pertains to the following:

- Name of injured employee
- Work activities the employee was performing within 24 hours of the onset of the injury or illness
- Description of injury or illness
- Formal diagnosis of the injury or illness and worker feedback on how the work evolution progressed
- Feedback from the employee as to what he or she believes caused the injury or illness

All companies have some form of information that is related to workplace injuries and illnesses, which is useful to the industrial hygienist in understanding and recognizing hazards in the workplace. In addition, company self-performed inspections or self-assessments are routinely conducted to verify workplace conditions and provide a record of conditions, as observed, in the workplace at the time of the inspection.

Depending on the industry, a self-assessment program can be very limited and look for those workplace conditions that only identify noncompliances with regulatory requirements, or the self-assessment programs can be formal and in depth and drive safety performance to not only address conditions that are regulatory noncompliances but also reinforce safe work behaviors. Typically, self-assessment programs address all topics of OSHA regulations and may include additional criteria that are important to the business and corporation. Trending of information generated from self-assessments can be useful in providing feedback to management and employees as to how existing processes and workplace conditions can be modified to promote safe work conditions.

Many companies have hazard identification checklists that are used in the hazard recognition process. Most of the checklists are focused on identifying all hazards posed to the worker and not just chemical or toxic materials. A checklist can be useful in stimulating thought from management and workers in recognizing hazards that either may not be obvious to them or they take for granted as being understood and recognized because they are common hazards. It is often the common hazards that pose the most significant risk to workers. Table 6.5 is an example checklist that can be used in the hazard recognition process. The checklist is generic to a general hazard recognition process but can be modified to focus on those hazards most often found in the industrial hygiene profession. Because Table 6.5 is a generic checklist, hazard recognition checklists should be tailored to the company and the hazards posed by the industry or work process.

6.6 CHEMICAL INVENTORIES: USE, STORAGE, AND DISPOSAL RECORDS

Both OSHA and the EPA require hazardous chemicals to be inventoried and managed to be protective of human health and the environment. Specifically, the United States Hazard Communication Standard, promulgated in 1983 and amended in 1987, requires hazardous properties of the chemical or compound to be readily identified on the container or package. The information has to be readily available to the workers for their use, and training on the hazardous properties of the chemical is required. The global harmonization system (GHS), as developed by the United Nations, was officially adopted in the United States by OSHA in 2009. The European Union (EU) Classification, Labelling, and Packaging (CLP) Regulation 1272/2008 adopted the GHS for all member states of the EU.[9] The United Kingdom retained the CLP regulation after they left the EU.

In the United States, all chemicals used in the workplace must be managed, and information communicated not only to the workers but also to local and state communities because of requirements associated with the Superfund Amendments and Reauthorization Act (SARA) Title III. SARA Title III, also known as the Emergency Planning and Community Right-to-Know Act (EPCRA) which was promulgated in 1986 following the accidental release of methylisocyanate in Bhopal, India, in 1984. As a result of the chemical release, more than 2,000 people were injured and chemical management regulations were promulgated that would require companies to inventory and provide chemical manufacturing and storage information to local emergency response personnel and county and state government agencies. Specifically, under EPCRA companies are required to:

TABLE 6.5
Example Checklist for the Hazard Recognition Process

Date: Walk-through location:__________________________________			
Person completing hazard recognition checklist:__________________________________			
	Yes	**No**	**Comments**
Chemical Hazards			
1. What work is performed in the work location that may use chemicals?			
2. Are health hazards of the chemicals readily known and is information available to the workers?			
3. Have there been any workplace incidents or illnesses identified for the chemicals used?			
4. Do workers know and understand the health hazards associated with the chemicals being used?			
5. Do workers understand how to control and manage exposure to the chemicals they use?			
Fire Hazards			
1. Are flammable products used in the work location?			
2. Are the flammable materials properly stored?			
3. Do workers understand the hazardous properties of the flammable material?			
4. What emergency measures are in place to manage a fire?			
5. Are the right type of fire extinguisher available?			
6. Are fire extinguishing and suppression systems routinely inspected?			
7. Do workers understand how to respond to a fire?			
Physical and Biological Hazards			
1. What are the physical and biological hazards of the work location?			
2. What are the physical and biological hazards recognized by the workers?			
3. Is noise generated as part of the work process?			
4. Are biological hazards, such as blood-borne pathogens, or unique biological hazards, such as bodily fluids or communicable diseases, present? (often associated with research and development activities)			
5. Do workers review work conditions prior to initiating work to ensure that any physical hazards have been identified?			
6. Do management and workers recognize that changes to their work tasks may trigger a need to review and confirm that no new hazards have been introduced into the work environment?			

- Provide material safety data sheets (MSDSs) for all chemicals used and stored at a facility
- Submit a toxic chemical release inventory form for all chemicals used above the applicable threshold quantities
- Notify regulatory agencies of releases greater than the reportable quantities (as defined in the regulation)

Chemical inventories, along with use, storage, and disposal records, can provide a piece of the puzzle for the industrial hygienist to use in understanding what health hazards may be posed when employees are performing work. A typical chemical inventory list contains the following information:

- Chemical name
- Chemical Abstract Services number
- Trade name
- Hazard class
- Manufacturer
- Quantity

Based on the information available, the industrial hygienist can then research and identify health hazard properties of the chemicals, as well as understand how the chemicals are used, to quantify health risk to the workers from exposure to the chemicals.

If more than one chemical is used in a work process, the industrial hygienist must also understand the additive and possible synergistic health effects that can be posed by more than one chemical. The use of more than one chemical by the employee can increase and further magnify the health risks posed by the chemicals and may drive additional administrative controls (such as lower acceptable occupational exposure limits) to be fully protective of the worker.

Storage of chemicals can also pose health risks to the workers. As workers perform their assigned tasks and increase the amount of time spent in a chemical storage area, there is an increased probability that they may be more frequently in contact with chemicals (and therefore potentially have an increased health risk) than those employees that just work with them. Documentation associated with the disposal of chemicals can also provide useful information to the industrial hygienist of what chemicals may be present as part of the manufacturing process.

The hazardous waste manifest and shipping information provide information associated with chemical properties and appropriate response protocol. The industrial hygienist can gain very useful information pertaining to the hazard identification and mitigation process when reviewing shipping information associated with the disposal of chemicals. State and federal government agencies will not allow anything to be shipped without specific information being identified, such as the chemical name and physical and health hazard properties. In addition, hazardous or chemical waste that is shipped for final disposal will not be accepted at the receiving facility without ensuring that the waste is compatible with the waste currently being stored and disposed of, and that the waste meets specific acceptance criteria, which can include toxicological considerations.

6.7 BIOLOGICAL AND RADIOLOGICAL HAZARDS CONSIDERATIONS

Most companies today have been required, out of necessity, to develop workplace policies on how to manage communicable biological hazards, such as Methicillin-resistant *Staphylococcus aureus* (MRSA) and Coronavirus-19 (COVID-19). In the United States, communicable illnesses used to be considered a personal illness and non-work related, but when a worldwide pandemic was declared with COVID-19, the Occupational Safety and Health Administration (OSHA) determined exposure to COVID-19 as work-related.

The reality of designating exposure to the virus as work-related meant companies now had to manage exposure to COVID-19 as a workplace hazard (evaluated as part of the work hazard identification process), but also they were financially responsible for health care costs associated with the exposure. During the height of the pandemic, companies were required to pay hospital and long-term disability costs because of the work-related designation. Similarly, exposure to radiation (both ionizing and non-ionizing) in the workplace must also be understood and controlled. Many industries use different devices, such as microwaves, laser devices, and non-destructive testing devices which emit radiation, and the industrial hygienist is responsible to understand the hazards associated with work equipment so all hazards can be identified and controlled.

6.8 REGULATORY INSPECTIONS AND VIOLATIONS

All companies will undergo some form of regulatory inspections. Regulatory inspections can be either planned or unplanned. Planned regulatory inspections are generally driven by regulatory agency criteria, such as number of employees, type of work being performed, or past performance and issues. Planned regulatory inspections may or may not be conducted by the same inspector each time, and usually include some level of follow-up after the identification of an issue or hazard that may be noncompliant. The industrial hygienist needs to understand what areas or operations have been inspected, along with the inspection results over the past five years (at a minimum), how the inspection results were managed, and whether the hazard still exists in the workplace.

If a noncompliant condition was cited, the noncompliance must be conspicuously posted and effectively communicated to the workforce. Because of regulatory liability, it is important that the industrial hygienist or safety professional review inspection results with the workforce, and it is recommended that workers be involved in the resolution process. Unplanned inspections are generally driven by either historical or poor safety performance or complaints from the workforce.

When responding to an unplanned inspection, it is important for one person to be designated the primary point of contact for the inspector, and that the company communicates when someone will be performing inspections on work being performed. The industrial hygienist should communicate how the work is performed, how hazards associated with the work are identified, and the process used to manage or mitigate the hazard. Historical inspection results, whether resulting from a planned or unplanned regulatory inspection, can often provide clues to the hazard identification process and can be useful to the industrial hygienist.

6.9 HAZARD CONTROL AND WORK EXECUTION

The industrial hygienist plays a major role in controlling hazards for the safe execution of work. Not only is the control of hazards an effective method for reducing the probability and severity of a health risk, but it is also regulatory driven by federal, state, and local agencies. As such, it is important that the industrial hygienist has a keen understanding of the model most commonly applied in defining the hierarchy of hazard controls. The most commonly recognized model for applying such a hierarchy is depicted in Figure 6.3.

Use of this model has proven to be an effective method to achieve a reduction in health risk and to protect the health and well-being of workers. In addition, by applying a hierarchy of hazard controls, both the behavior of the workers and the overall environment in which work is performed are improved and result in a greater acceptance of risk. Using the hierarchy of control model, the first method for hazard control should begin with an attempt to eliminate the hazard, and the method that should be considered last is the use of personal protective equipment (PPE).

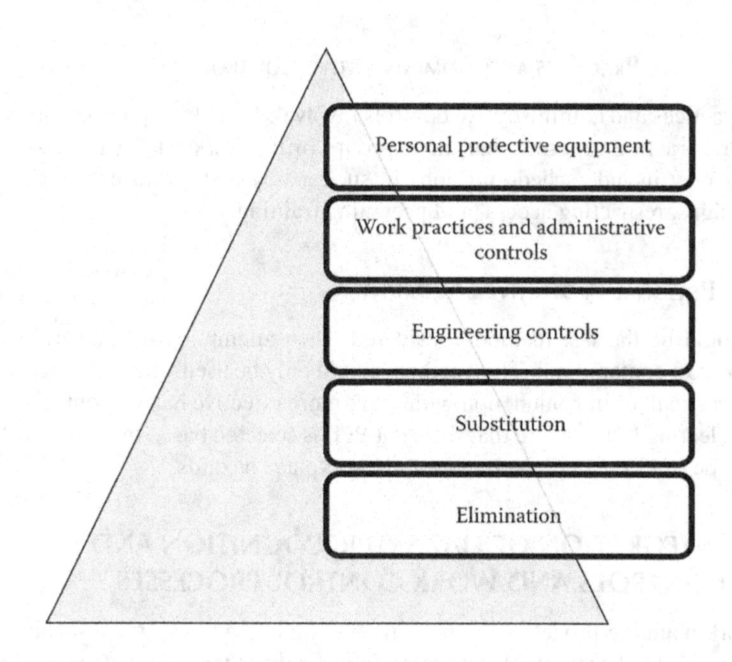

FIGURE 6.3 Hierarchy of hazard controls model.

6.9.1 Hazard Elimination

The most effective method to control a hazard is to completely remove it from a work or manufacturing process. Elimination of the hazard may mean that the entire process is revised to change the operating parameters to reduce the interface between the process or machines and the worker, for example, changing a process to remove the requirement to use a known carcinogen. This method is most easily applied in the design of a system before equipment is purchased and installed.

6.9.2 Product Substitution

Product substitution provides an opportunity to replace a highly hazardous substance with a less hazardous one, or a hazardous substance with a non-hazardous one. In the former case, a hazard may still exist and further hazard control methods may be needed.

6.9.3 Engineering Controls

Engineering controls involve designing and adding physical safety features and barriers to the process or equipment to eliminate or reduce exposure of the worker to hazards. Examples of engineering control include adding a wet scrubber to a process to reduce dust emissions, adding ventilation to eliminate or reduce the release of fumes from a chemical process, or adding a guard to a table saw to prevent worker contact with the blade during operation.

6.9.4 Work Practices and Administrative Controls

Work practices and administrative controls involve establishing policies, procedures, and work practices to minimize the exposure of the worker to health risks. These controls may include scheduling jobs in such a way as to limit exposure, posting hazard signs, restricting access, and providing training.

6.9.5 Personal Protective Equipment

PPE should be the last method considered when attempting to reduce a worker's exposure to a contaminant. This method should only be used when all other methods are impractical, or in conjunction with other more effective hazard control methods. When selecting PPE, ensure that the right PPE is selected based on the hazard, as no single type of PPE can protect against all workplace hazards.

6.10 INTEGRATION OF HAZARD RECOGNITION AND CONTROLS AND WORK CONTROL PROCESSES

The work planning process is at the core of completing work and ensuring that the work is performed on schedule and safely, and results in the quality desired. Planning

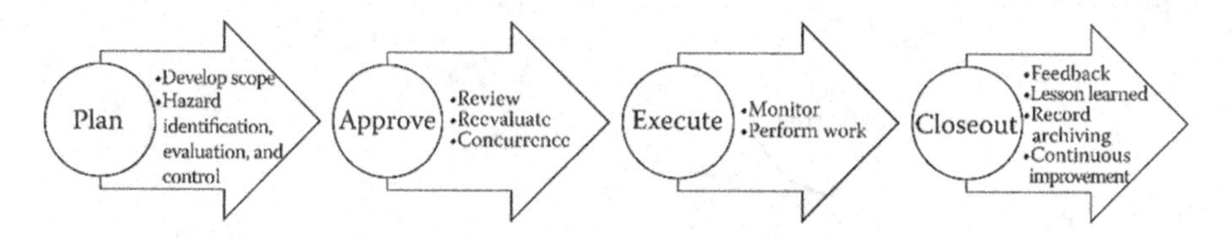

FIGURE 6.4 Work planning and control guidance.

work from beginning to end, and integrating the hazard recognition and control process, is key for company success and requires a concentrated effort to ensure that each aspect of the job is reviewed and analyzed, and health hazards and associated risks are mitigated when discovered. There are four important stages that are critical in the work control process: planning, approving, or authorizing, execution, and closeout. Figure 6.4 depicts a guide for planning and controlling the flow of work.

6.10.1 Planning Work

A work plan is the first stage of the process used to control the manner in which work is performed. This plan provides a comprehensive account of how a specific task will be accomplished. The work planning process includes:

- The ability to develop and bound the scope of the work
- Performing a comprehensive analysis of work tasks or steps
- Identifying and mitigating hazards associated with each work task or step

A properly documented and bounded work scope is important in the hazard analysis process. Often, hazards are missed during the evaluation process because the scope is not bounded properly, allowing the potential for scope creep to manifest itself. Scope creep refers to the process of allowing a project or task to grow beyond the intended purpose, scope, or size. Planning activities that can assist the industrial hygienist in the work planning process include:

- Developing a clear scope. Figure 6.5 identifies three attributes to consider when defining the work scope.
- Ensuring that hazards are evaluated and mitigated. Figure 6.6 identifies three attributes that should be considered when evaluating and mitigating hazards.
- Involving the worker in the scope development and hazard analysis process.
- Engaging the appropriate subject matter experts.
- Reviewing the scope to ensure that it is appropriately bounded.

6.10.2 Authorizing Work

The project manager or responsible individual must be confident that work is ready to proceed safely before approving work to begin. This is one of the greatest

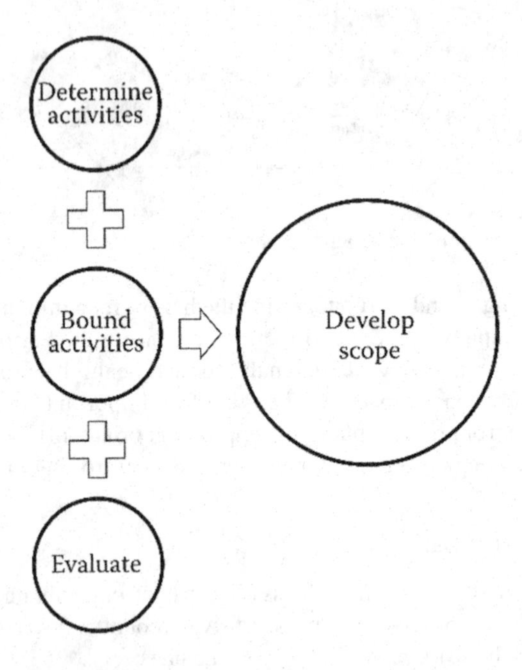

FIGURE 6.5 Development of project scope.

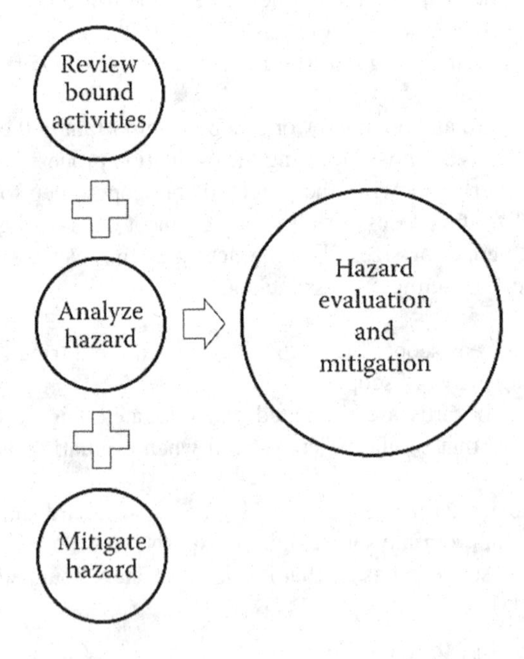

FIGURE 6.6 Hazard evaluation and mitigation.

responsibilities that has been entrusted to project managers and supervisors since they are responsible for facilitating a safe work environment for workers. Before approving work, there must be a concentrated effort to ensure that all parties involved are in agreement that the project or task is ready to be worked. This includes considerations of:

- Ensuring the scope is appropriately bounded. This is a good time to review and reevaluate the scope.
- Hazards being identified, analyzed, and mitigated.
- Tools and resources being available to complete the job safely.

Once the project manager and supervisor are comfortable that all project parameters have been considered and the right measures are in place, then final consent can be given to begin work. In addition, it may be necessary to receive consent from

- Subject matter experts
- Supervisors
- Project management
- Workers (final worker consent is generally given at a tailgate or pre-job meeting before work can begin)

6.10.3 Work Execution

Work should be executed as planned using the approved scope of work. Supervisors and workers must avoid the temptation to include other task in the process that have not been evaluated and approved. It is important to have the scope clearly defined to keep this from happening. Before beginning the work, workers must be briefed on the following to ensure that work quality and safety are at the forefront of work:

- Scope of the work
- Potential hazards and mitigating controls
- Protective equipment that will be required for the job
- The right of workers to pause work when they believe safety is of concern or the project is not being implemented as planned

Workers must also be empowered to call a work pause when things are not going as planned, when there is a potential hazard that has not been mitigated, or if they are unsure how to carry out a particular step. Supervisors and management must monitor the work to ensure that it is being performed as planned and safely. Figure 6.7 identifies the workflow process the industrial hygienist uses when performing or supporting the execution of work scope.

6.10.4 Project Closeout

During a project closeout, it is time to evaluate how the project performed and to learn from project successes and failures. Figure 6.8 depicts attributes the industrial

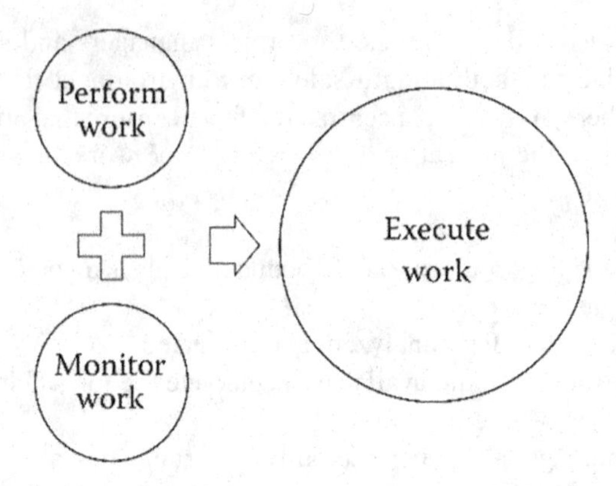

FIGURE 6.7 Work execution process.

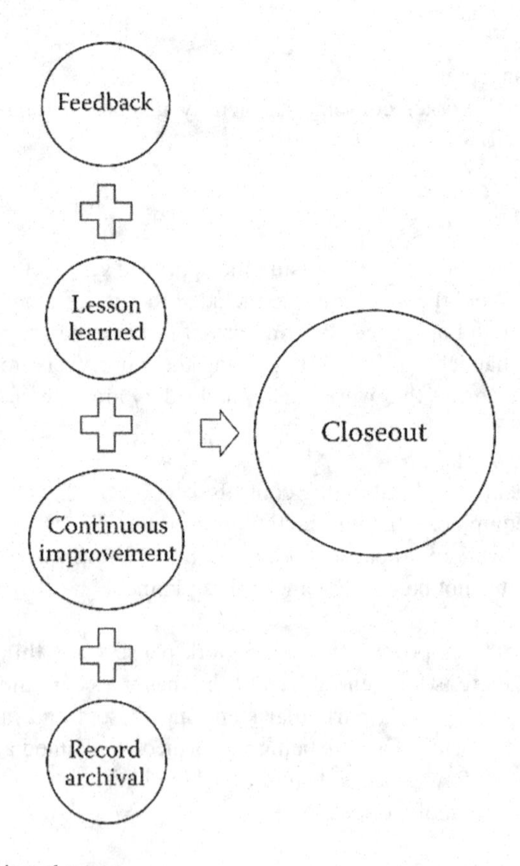

FIGURE 6.8 Project closeout.

hygienist should consider when supporting project closeout activities. It is also a good time to determine what can be done better in the future to improve work implementation and quality and reduce health risks to the worker. As an industrial hygienist, you should reflect on how health hazards were identified, the methods and approaches for evaluating those health risks, and how hazard controls were applied to identify improved methods for reducing those health risks. Incorporating the worker into this self-reflective process is not only beneficial to industrial hygienists in improving their skills and the reduction of health risks, but also empowering to the workers because they are a part of building a better safety and health program.

6.11 MANAGEMENT BY WALK-AROUND

The focus of management by walk-around should be on observing work behaviors and reinforcing management expectations of having a safe work environment. Management spending time in the field or on the shop floor is a great way to become aware of potential workplace hazards, as well as build trust and relationships with the workers. Workers will open up to and share information with managers that they believe are concerned with their well-being and the way the business operates. There are three key elements that make this style of management successful: preparation, communication, and persistence to connect. Figure 6.9 depicts these key elements.

6.11.1 Preparation

It is necessary that the manager prepares to engage in meaningful and insightful conversations and interactions with workers as he or she is walking through the work area. It may be necessary for managers to familiarize themselves with the types of projects and tasks that workers are performing before interacting with them. Workers are generally always happy to discuss their contribution to the company and are even

FIGURE 6.9 Management by walk-around paradigm.

more excited to discuss their jobs when the leader shows that he or she has some level of familiarization with the tasks and business. In addition, it would be helpful if the leader has some familiarization with the workers and their performance.

6.11.2 Communication

Communication with workers must be approached openly, honestly, and with sincerity. Workers will know when a manager is not being forthright and sincere. Therefore, when management and professionals are walking around, there should also be sensitivity regarding how to focus on identifying and correcting unsafe behaviors.

6.11.3 Persistence to Connect

It is entirely possible that during initial walk-arounds, employees will be reluctant to speak with the manager. In such cases, that manager must be prepared to find an area of interest that will spark interest and energize the worker into engaging in meaningful dialogue. It may take several trips to the work area to spark communication from some workers. Some workers will need to see consistency in actions from leaders before opening up and offering ideas and suggestions.

6.12 SAFETY THROUGH DESIGN: DESIGNING HAZARDS OUT OF THE PROCESS

The most effective and efficient method to mitigate workplace hazards, and associated health risks, is safety through design during the conceptual phase of a project or task. Safety through design is a concept gaining importance that refers to the integration of hazard identification and control methods in the design of processes and systems. Safety through design also refers to the incorporation of risk assessment methods early in the design process to eliminate or minimize the risks of injury to workers. Figure 6.10 depicts items to be considered, by the industrial hygienist, when performing reviews or involved in the overall design process. An effective design strategy should encompass all design activities, including facilities, hardware, systems, equipment, processes, and products.

In order to evaluate and manage safety and health risks during the design phase, management, subject matter experts, and engineers should consider the following:

- Physical design of a product, process, or system
- Workplace layout, with a focus on eliminating or reducing potential hazards
- Factoring controls in the design process to eliminate hazards that may occur during operation or processing
- Designing work, workstations, and processes to eliminate or minimize the risks to workers

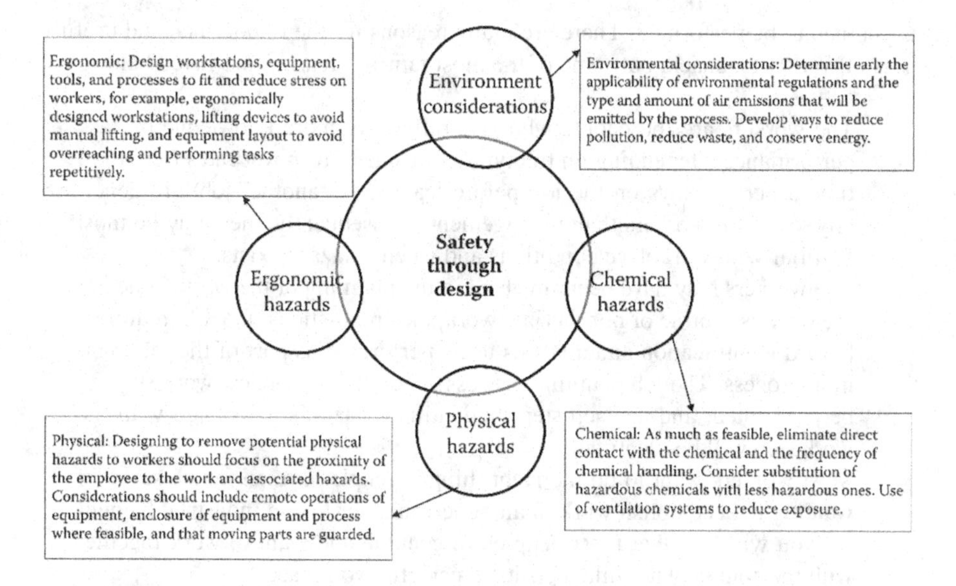

FIGURE 6.10 Safety through design considerations.

Many of the environment, health, and safety regulations have several requirements that may be more effectively addressed through the design stage of a process. Some regulatory requirements that can be effectively designed in the design stage include provisions for hazards such as the following:

- Noise
- Confined space
- Hazardous chemicals
- Stack discharges

Ensuring that workstations, work processes, and systems are designed to reduce or eliminate the risk to workers is an important part of a worker health and safety management system and management's commitment to their workers.

Waiting to address process hazards at the completion of the project can be a costly and less effective way to address safety and health issues, and at times, it may be too late to eliminate or minimize hazards. At such points, oftentimes workers are asked to wear PPE that can at times present additional hazards or stress on the worker when performing work tasks.

6.13 EMPLOYEE ENGAGEMENT AND INVOLVEMENT

One of the primary mechanisms for identifying, evaluating, and controlling workplace hazards is feedback from the workers. The workers are the industrial hygienist's eyes and ears in the work environment and are most familiar with the work that is planned to be performed. There are many reasons for the importance and usefulness of worker feedback, but some of the most familiar reasons are listed below:

- The workers are the people who generally have more knowledge on task performance. Depending on the type of business, turnover rates (length of time a person stays on the job before leaving for another job) are generally lower for workers than management. Consequently, they may be most familiar with workplace conditions and when a hazard exists.
- The workers may have been involved in the planning of the work. Whether the work is routine or nonroutine, workplace regulations (OSHA) require a hazard identification and analysis to be performed as part of the job planning process. The job planning process generally will review work steps to be performed, and as each step is identified, hazards associated with the work step will be identified.
- Most workers want to do the right thing – people have an innate sense of wanting to accomplish work in an honest manner. As an industrial hygienist, you will find that most people are genuine and want to work together with the company to build a positive, prideful workplace.
- Workers tend to trust fellow workers more than company personnel. Relationships are built over time among fellow workers because they spend a significant amount of time with each other. In many cases, work crews become like a family. Therefore, when workers identify hazards, they may be more readily accepted and managed by fellow workers.

Identifying a hazard may not always be viewed as a positive behavior trait by some within the company, so the industrial hygienist should recognize that the raising of an issue or identification of a hazard should be encouraged. In addition, a good industrial hygienist should work on building a relationship with the workforce in order to increase trust, communication, and the ability to work together in improving workplace conditions. It should also be noted that most inspection and regulatory personnel will interview workers when performing an evaluation or inspection; therefore, it is prudent for the industrial hygienist to have a good relationship with the worker to ensure that information is being communicated on a daily basis so that the company may have an opportunity to address an issue prior to it being communicated outside the company.

Most would agree that a key element in providing a safe workplace begins with employees that are actively engaged in work and the work environment. When employees are engaged in hazard evaluation and mitigation, they are invested in ensuring that the project and the associated tasks are performed within the desired quality and safety envelope. Some of the benefits of employee engagement include:

- A willingness to work to make the company successful
- A higher level of trust
- Innovative thinking
- A commitment to the safety culture of the company

Employees who are engaged in their workplace are a known asset in their ability and willingness to seek out hazards and work with leadership in evaluating and mitigating those hazards. To ensure that employees are equipped to assist effectively with hazard recognition and mitigation, it is incumbent upon management to ensure that the appropriate training is provided. Training employees on the process of hazard recognition is extremely important. Workers are close to the work that takes place in an organization and company. Hazard recognition training for workers is discussed in more detail in Chapter 8.

Questions to Ponder for Learning

1. Discuss the importance of relationship building for the industrial hygienist.
2. Name and discuss some benefits of employee engagement.
3. What are some of the chemical and physical hazards an industrial hygienist must mitigate in the workplace?
4. Why is it important for an industrial hygienist to be proficient in the recognition, evaluation, and control of workplace hazards?
5. What are some tools industrial hygienists can use to guide them in the identification of workplace hazards?
6. Discuss the right of an employee to refuse to perform a task that he or she believes to be unsafe and the actions employers can take to address the situation.
7. When should hazard recognition training be administered?
8. Discuss some poor work practices and human behaviors that can facilitate and contribute to unsafe work practices and resultant increased health risks.
9. How can a hazard identification checklist aid the industrial hygienist in recognizing hazards that are present in the workplace?
10. What is a common model used for identifying hazard controls, and how should the industrial hygienist implement the model in the decision-making process?
11. List and discuss the stages of the work planning process.
12. What is the role of an industrial hygienist in the work planning and control process?
13. Define *scope creep*. Discuss how it can introduce hazards that were not anticipated and mitigated.
14. Describe the safety by design process. How important is this concept in controlling hazards in the workplace and ultimately reducing health risks?

REFERENCES

1. U.S. Chemical Safety and Hazard Investigation Board (CSB), website: http://www.csb.gov
2. National Transportation Safety Board, 2023. *Norfolk Southern Railway Train Derailment and Subsequent Hazardous Material Release and Fires.* Investigation Details. https://www.ntsb.gov/investigations/Pages/RRD23MR005.aspx

3. Tennessee Valley Authority, *Regulatory Submittal for Kingston Fossil Plant.* https://www.documentcloud.org/documents/20423066-on-scene-coordinator-report-for-the-time-critical-removal-action#document/p118
4. U.S. Environmental Protection Agency, 2023. *EPA Response to Kingston TVA Coal Ash Spill.* https://www.epa.gov/tn/epa-response-kingston-tva-coal-ash-spill
5. The Guardian, 2023, February 25. *Revealed: The US is Averaging One Chemical Accident Every Two Days.* https://www.theguardian.com/us-news/2023/feb/25/revealed-us-chemical-accidents-one-every-two-days-average
6. Coalition to Prevent Chemical Disasters, https://preventchemicaldisasters.org/
7. *Reporting of Injuries, Diseases, and Dangerous Occurred Regulations* RIDDOR. https://www.legislation.gov.uk/uksi/2013/1471/contents/made
8. OSHwiki, *Reporting and Monitoring Occupational Accidents and Diseases in Europe.* https://oshwiki.osha.europa.eu/en/themes/reporting-and-monitoring-occupational-accidents-and-diseases-europe#:~:text=In%20the%20European%20Union%2C%20there%20is%20a%20legal,diseases%20and%20other%20work-related%20health%20problems%20and%20illnesses
9. EUR-Lex. *Access to European Law.* https://eur-lex.europa.eu/legal-content/EN/LSU/?uri=CELEX%3A32008R1272#:~:text=It%20lays%20down%20uniform%20requirements%20for%20the%20classification%2C,chemicals%20appropriately%20before%20placing%20them%20on%20the%20market

7 Medical Monitoring and Surveillance of the Worker

7.1 INTRODUCTION

Medical monitoring and surveillance of the worker encompass both the medical techniques and the methods used to evaluate workers exposed to physical, biological, chemical, and radiological hazards, along with a specified frequency and program that validates that the workplace environment does not cause harm to workers.

A medical monitoring and surveillance program is an essential element of an overall worker injury and illness protection and prevention program. Most companies and institutions have, and maintain, a medical monitoring program for their workforce. The benefits of a medical monitoring and surveillance program include, but are not limited to:

- Ensuring that personnel are capable and able to perform a job prior to being hired
- Reducing the cost of health insurance provided by the company
- Reducing workers' compensation costs covered by the company
- Confirming to employees that the work they are performing does not impact their personal health and demonstrating a reduction in health risk
- Providing employees with access to wellness and injury illness prevention information
- Confirming to a company or institution that its hazard controls are successful and are not causing additional harm to the employees
- Complying with regulations that require the workplace to have a medical monitoring and surveillance program

The industrial hygienist is a key player in the effective implementation of a medical monitoring and surveillance program. Defining the type of hazards that the worker is exposed to and locations of the hazards, anticipating and preventing negative health effects due to exposure to the hazards and contaminants, and preventing an increased health risk to the worker are just some of the reasons why the industrial hygienist is key to ensuring that health risks to workers are minimized.

A medical monitoring and surveillance program should be tailored to the hazards and contaminants to which the worker may be exposed. Occupational health

DOI: 10.1201/9781032645902-7

medical providers are not usually educated in workplace hazards because their primary means of evaluating patients is by symptoms and body functions' actions and reactions. They may or may not have ever visited the work location of their patients, and many times their services are contracted, with little information provided to the medical provider so they can better understand the work environment and the potential for causation of an injury or illness. It is often the responsibility of the industrial hygienist to work with the medical provider to ensure that a good communication mechanism exists to bridge the informational gap so that the medical provider can make the best determination on whether an injury or illness was caused by the workplace, because of actions taken by the worker while performing their job. The occupational health medical provider also has access to a worker's personal medical history and will use that information in determining "causation."

Much of the information used by the industrial hygienist when interfacing with the occupational health medical provider is related to regulatory requirements. In the United States, there are common regulatory requirements associated with medical monitoring that are cross-cutting and applicable to all industries, such as requirements associated with asbestos, lead, chromium, noise, and heat stress. There are also regulatory requirements associated with medical monitoring that are specific to the type of profession, such as nuclear operations and remediation, hazardous waste site cleanup, and chemical demilitarization. It is the responsibility of industrial hygienists to investigate what medical monitoring requirements are applicable to the work they are professionally supporting and incorporate those requirements into the company's occupational safety and medical programs. Additionally, information generated as part of the medical monitoring and surveillance program may be used to support workers' compensation claims. The industrial hygienist is the primary person relied on for managing and providing exposure information, such as air monitoring and sampling, surface sampling of particulates, and general exposure information that may be needed to process a workers' compensation claim.

When a worker is injured or has exhibited symptoms that could be attributed to an occupational illness, it is the industrial hygienist who is relied on to provide workplace monitoring information that can either support or refute the claim. Workers' compensation programs are administrated by each state, so depending on where the company, industry, or institution is located, the type of information needed to support the program needs to be understood by the industrial hygienist. All these factors must be considered by industrial hygienists when they are establishing and managing their workplace medical monitoring and surveillance programs.

7.2 MEDICAL MONITORING AND SURVEILLANCE PROGRAM

Almost every health and safety regulatory body that exists in the United States, and worldwide, requires some form of medical monitoring for workplaces that fall under its jurisdiction. The number of requirements that may apply depends largely on the work environment and hazards that are present and formed in the course of performing work. For example, in the United States the Occupational Safety and Health Administration (OSHA) requires a worker who is exposed to inorganic arsenic at levels greater than 5 $\mu g/m^3$ more than 30 days per year to be annually examined.

Another example is that an employee who performs work at a hazardous waste site is required to obtain a medical exam annually or when there is a significant change in a work assignment (being introduced to a new regulated contaminant or hazard). Similar medical surveillance requirements are imposed by other countries and agencies for exposure to various constituents in the workplace. One such agencies include the Health and Safety Executive, Britain's regulator of health and safety in the workplace.

It is the role of the industrial hygienist to be familiar with the work activities, workplace hazards, and medical monitoring requirements associated with the work being performed. Depending on the size of the company, the medical monitoring and surveillance program may be managed within the human resources department (as with smaller companies), within the safety and health group (midsize groups), or by departments that employ the workers (for larger companies). The smaller the company, the more involved the industrial hygienist may be with the occupational health provider. There are essentially six elements of a medical monitoring and surveillance program:

1. Establishment of company protocols related to the medical monitoring and surveillance program.
2. Scheduling and tracking of medical physicals.
3. Performance of the medical physical, whether it be a routine visit or emergency.
4. Notification of results, whether from the occupational health provider or workplace sampling and monitoring.
5. Analysis of occupational health data and sampling and monitoring results.
6. Records management.

To some extent, industrial hygienists are involved with each of the elements. The degree of their involvement varies depending on whether the program element is directly related to the hazard analysis and control process. Those elements of the program that implement administrative mechanisms, such as scheduling and tracking of physicals, and records management, require less involvement of the industrial hygienist; however, those programs that are integral to the diagnosis of an injury or illness typically require an increased level of involvement because of the realization of a health risk. Table 7.1 depicts each of these elements and the role the industrial hygienist plays in the implementation of an effective program.

7.3 ESTABLISHMENT OF COMPANY POLICIES, PROTOCOLS, AND PROCEDURES

The foundation of a medical surveillance and monitoring program is based on company policies, protocols, and procedures. Almost every company or institution has an established policy that forms the basis of protocols and procedures to be followed when executing the program. In addition, the company policies establish the commitment of the company or institution to protect the worker. The policy will vary depending on the hazards to be encountered. For example, if the company or

TABLE 7.1
Elements of a Medical Monitoring and Surveillance Program

Medical Monitoring and Surveillance Program Elements	Industrial Hygiene Involvement
Establishment of company protocols	• Development of policies and procedures related to preemployment, daily work and emergencies, fitness for duty, and postemployment physicals
Scheduling and tracking of physicals	• Identification of required medical tests—driven by regulations and company policy • Generally implemented via commercially available software
Performance of medical physical	• Identification of contaminants the worker is exposed to • Determination of frequency of performing workplace tasks • Identification of workplace environmental conditions
Notification of test results	• Communication of an injury or illness diagnosis to management and outside stakeholders (if required) • Primary interface with the occupational health medical provider
Analysis of occupational health data	• Evaluate occupational health data (generically), along with workplace sampling and monitoring data
Records management	• Documentation associated with any notification of a potential injury or illness of a worker; may include restrictions and documentation of management to minimize further injury

institution is focused on biomedical research and development, then the policy would be focused on the work categories that would be potentially exposed to the hazard and identification of the biological agents. An appropriate medical surveillance and monitoring policy generally includes the following criteria or expectations:

- A written program that depicts the commitment by the company to ensure a safe work environment
- Establishment of a program that identifies and mitigates hazards
- Training to be provided to employees to ensure that they are knowledgeable of the hazards to be encountered and associated health risks, as well as controls to mitigate injury or illness
- Expectation of workers to report all injuries and illnesses
- Commitment to provide treatment if an injury or illness should occur
- Additional services the company will provide should an injury or illness occur (e.g., counseling or insurance assistance)
- Commitment to work with the union, as part of the medical monitoring and surveillance program, if applicable

Upon issuance of a company policy, the company or institution should develop protocols and procedures for implementing the surveillance program. Company protocols identify criteria associated with the surveillance program, and procedures integrate specific protocols to be followed. Many large corporations establish medical response and surveillance protocols that are integrated into procedures used by different organizations within the company to implement the medical monitoring and surveillance program.

7.4 SCHEDULING AND TRACKING OF PHYSICALS

The scheduling and tracking of employee physicals are generally not the responsibility of the industrial hygienist. However, depending on the size of the company, the industrial hygienist may be responsible for providing safety and health services, along with keeping track of all workers and their required physicals. At a minimum, the industrial hygienist should understand and review the scheduling and tracking mechanism used to ensure that the company is meeting regulatory requirements associated with worker qualification and confirm that appropriate scheduling and tracking of physicals are occurring. The frequency with which the physicals must occur is dependent on the specific contaminant the employee is exposed to and, in some cases, the concentration of the exposure. For example, in the United States the occupational health standard for inorganic lead dictates that a medical surveillance program must be made available to all employees exposed to lead above an action level of 30 $\mu g/m^3$ time-weighted average for more than 30 days each year. Under this program, the blood lead level of all employees who are exposed to lead above the action level is to be evaluated at least every six months. This frequency is increased to every two months for employees whose blood lead level is between 40 µg/100 g and the level requiring employee medical removal. For employees who are removed from exposure to lead due to an elevated blood level, their blood lead level must be monitored monthly; therefore, even if an employee is no longer performing the job that caused the exposure, he or she may still need to be in a medical surveillance program until such time their blood levels decrease to acceptable values. It is also worthy to note that OSHA requires a worker to be placed into a medical removal protection program if their blood lead levels are substantially elevated or otherwise at risk of sustaining material health impairment from continued substantial exposure to lead. The industrial hygienist should be familiar with all job categories within the company or institution and understand the different mechanisms through which an employee could be exposed to physical, chemical, or biological contaminants, and the medical requirements associated with those job categories.

Some companies use an employee work assignment analysis to assist in identifying essential job functions and the associated physical or chemical hazards. An employee work assignment analysis can be generated when an employee is hired to document anticipated environmental and workplace conditions to be encountered, and it is provided to the occupational health medical provider as a background document for the establishment of a medical surveillance program for a worker.

The employee work assignment analysis can also be used to initially assign a worker to a similar exposure group (SEG), which the industrial hygienist uses to identify and manage potential occupational exposures in the workplace. An example employee work assignment analysis form is provided in Table 7.2. The form is very

TABLE 7.2
Example Employee Work Assignment Analysis Form

Employee Work Assignment Analysis:

Page 1

Date: 01/01/2017
Employee name: Joe Smith
Job title: Industrial hygienist
Assigned manager: Candy Cane

Work Assignment	**Assigned SEG**
• Daily office work with computers	1
• Perform industrial hygiene instrumentation daily checks and calibration	1
• Perform sampling and monitoring in the field	3
• Attend meetings	1
• Respond to emergencies in the field	4
Physical Job Activities	
Average time spent sitting/day?	<4 hours/day
Average time spent walking and standing/day?	<4 hours/day
How often is climbing involved in the job?	<1 hour/day
Average time on computer/day?	<4 hours/day
Average time spent operating a motor vehicle/day?	<2 hours/day
Amount of time spent handling material and equipment?	<1 hour/day
Are work activities performed in hot or cold environments?	Yes. The employee may perform industrial hygiene sampling and monitoring in both hot and cold environments for <10 hours/week
Will the employee be required to lift objects greater than 10 lb?	No. General job duties will require the employee to lift calibration and instrumentation equipment, along with stretching when performing monitoring, but not expected to be >10 lb
Are there repetitive activities the employee may be performing?	Yes. The employee will be doing computer work and programming industrial hygiene instruments
Will the employee be required to kneel as part of their job functions?	Yes. The employee may be required to kneel when performing sampling and monitoring activities
Will the employee be using tools that vibrate?	No. The employee will not be using vibrating tools

(*Continued*)

TABLE 7.2 (CONTINUED)
Example Employee Work Assignment Analysis Form

Employee Work Assignment Analysis:

Page 2

Date: 01/01/2017
Employee name: Joe Smith
Job title: Industrial hygienist
Assigned manager: Candy Cane

Required Work Qualifications

1 Wear a respirator wearer
2 Work with hazardous waste
3 Chemical management
4 Emergency response
5 Work with asbestos
6 Work with beryllium
7 Work with lead
8 Work within a confined space

Potential Exposures to Contaminants

Contaminant	Frequency of Exposure
Asbestos	<40 hours/year
Lead	<10 hours/year
Beryllium	<10 hours/year
Noise	<40 hours/year greater than 85 dB
Mercury	<10 hours/year
Benzene	<40 hours/year
Chromium (hexavalent)	<40 hours/year
Chlorinated solvents	<80 hours/year
Isocyanates	<10 hours/year
Ammonia	<10 hours/year
Nitrosodiethylamine	<10 hours/year
Ionizing radiation	<10 hours/year
Nonionizing radiation	<40 hours/year
Printer toner	<5 hours/year
Paints and solvents	<10 hours/year

useful in providing a vehicle to communicate with the occupational health medical provider about anticipated exposures and related medical surveillance programs, such as hearing conservation, asbestos work, lead work, and confined spaces. In the United States, the form is also useful in managing compliance with the Americans with Disabilities Act (ADA) and fitness-for-duty issues.

7.5 INTERFACING WITH MEDICAL PROFESSIONALS

It is the role of the industrial hygienist to frequently interface with the designated occupational health medical provider. The building of a close working relationship is extremely important because the occupational health medical provider can assist the company in managing the injury or illness, as well as decrease the amount of time required to return an employee back to work and prevent reaggravation of an injury. In addition, returning the employee back to work sooner improves their mental health because every person has an internal desire to do a good job, while positively contributing to supporting their family and the company.

The industrial hygienist also plays a critical role in educating the occupational health medical provider on workplace hazards. The medical provider may not understand what tasks must be performed when a worker is employed by a company, and he or she most likely will not know what hazards the worker may be exposed to in the performance of their job. Some job categories may only expose the worker to hazards that present a low health risk, such as an administrative assistant, while other jobs may expose workers to unknown hazards that pose more significant health risks, which are frequently encountered when performing work at hazardous waste sites.

The industrial hygienist is relied upon to communicate any environmental conditions that are present in the workplace, such as heat, cold, physical terrain limitations, and radon background levels, which could also influence how the occupational health medical provider would approach implementation of a medical surveillance and monitoring program. Depending on the occupational health medical provider, they may be interested in attending a tour of the facility to gain a better understanding of the work activities and associated medical health risks. In the United States, the U.S. Department of Energy requires contractors to provide occupational health medical providers access to the workplace for evaluation of job conditions and issues relating to workers' health.[1]

The industrial hygienist will correlate the SEGs with the job categories; this information is then communicated to the occupational health medical provider for use in performing medical physicals. The industrial hygienist is also relied upon by the occupational health medical provider, to assist them understanding general management and health and safety terminology used by the company or institution. The industrial hygienist must understand the ability and effectiveness of the occupational medical provider to prevent, diagnose, and treat occupational illnesses and diseases directly correlated with the quality of data and information provided by the industrial hygienist. Data must include the quantitative and qualitative aspects of evaluating workplace exposures.

Establishing a routine interface meeting with the occupational health medical provider is always a good idea in establishing and building the relationship. Suggested topics to be discussed in these meetings include.

- Workplace injuries or illnesses that have occurred over the past month or six months

- Any specific type of injury or illness that the occupational health provider is observing to occur at an increased frequency
- Recommendations of the occupational health provider for implementing additional actions in the workplace to prevent future injuries and illnesses
- Any medical or insurance information needed to support workers being evaluated by the occupational health provider
- Actions that both the industrial hygienist and the occupational health medical provider can implement to strengthen the medical surveillance and monitoring program

It is the role of the industrial hygienist to be the eyes and ears of the occupational health medical provider when implementing a medical surveillance and monitoring program.

7.6 NOTIFICATION OF TEST RESULTS

Following the performance of the medical physical, the occupational health medical provider will notify the worker of the test results. The industrial hygienist provides support to the worker in helping them understand the correlation between their work activities and the results of the physical, particularly if the medical results are not favorable. The industrial hygienist, in conjunction with the occupational health medical provider, can help the worker understand the current health risk or impact, how to prevent further impact or injury at the workplace, and actions he or she can take to continue to minimize health impacts.

It is at this time that the industrial hygienist may be involved in working with the occupational health medical provider, and others within the company, such as the worker's supervisor and human resources, to remove or restrict the worker, either temporarily or permanently, from performing work tasks that could aggravate or worsen the medical diagnosis. The occupational health medical provider may also require additional medical tests to be performed to confirm a diagnosis and causation, and these tests could be a one-time series or ongoing, such as withdrawing blood every two months, as required in the lead medical surveillance and monitoring program.

7.7 ANALYSIS OF OCCUPATIONAL HEALTH DATA

Medical health data received from the occupational health medical provider can be very useful to the industrial hygienist in preventing and responding to workplace injuries and illnesses. Many companies subcontract their occupational medical services, so industrial hygienists may need to compile the information themselves, which is another reason why it is smart to establish a good working relationship with the occupational health medical provider. Below are some examples of information that is useful to the industrial hygienist in understanding how hazards in the workplace manifest themselves into workplace injuries and illness.

- The types of injuries or illnesses that have occurred over the past six months, one year, or two years
- Illnesses positively linked to work having been performed in the workplace
- Verbal feedback from workers as to what they believe caused their injury or illness

Based on information obtained from the medical surveillance and monitoring program, the industrial hygienist can evaluate existing program data, such as air monitoring results, surface smear results, and direct-reading instruments, to determine if hazard controls have been effective. Through evaluation of monitoring and sampling data, along with feedback from the occupational health provider and feedback based on medical examinations, the industrial hygienist can validate that no overexposure of workers to contaminants has occurred. For example, several workers in a similar exposure group (SEG) have been identified as having elevated lead concentrations in their blood based on the results of their most recent medical physical. Review of the air sampling data, along with personal sampling results, indicates that when workers perform a specific task using a particular type of metal, increased air concentrations occur and may be the cause of elevated lead levels in the blood. Blood work results from medically evaluating the rest of the work crew, who are not assigned to a particular SEG, indicate that they do not exhibit elevated blood levels. Consequently, the industrial hygienist will work with engineering and the occupational health medical provider to minimize additional exposure to lead and the cumulative health effects exhibited by exposure to the contaminant.

7.8 MEDICAL MONITORING RECORDS AND REPORTING

Medical records management may be driven by more than one regulation: those that regulate safety and health programs in the workplace, and medical standards that may apply to occupational health medical providers - depending on the industry. OSHA Publication 3110, published in 2001, provides a general overview of regulatory requirements that drive worker access to medical records. Personnel who have a right to access their relevant exposure and medical records include:

- Current or former workers who have been or are currently being exposed to toxic substances or harmful physical agents
- Workers who have been assigned or transferred to perform work that involves toxic substances or harmful physical agents
- A legal representative of a deceased or legally incapacitated worker who has or may have been exposed to toxic substances or harmful physical agents
- Designated employee representatives under specific circumstances, with written authorization

It is important to note that employers do not have to make all records available for review.

The industrial hygienist is often relied on to be the representative for the company when providing access and managing exposure and medical records. In the United States, the industrial hygienist should be aware that under OSHA, employers have the responsibility to

- Preserve and maintain accurate medical and exposure records for each employee
- Inform workers of the existence, location, and availability of those medical and exposure records
- Give employees any informational material regarding exposure and medical records that OSHA makes available
- Ensure that records are available to employees, their designated representatives, and OSHA, as required

If, for some reason, exposure records for an employee are not available that document the amount of a toxic substance or harmful physical agent, the company is required to provide records of other workers who performed similar job tasks or performed tasks in a similar work environment as the employee requesting the record. The company may also be required to provide exposure records that reasonably indicate the amount and nature of toxic contaminants at a particular workplace, or used in a specific working environment, in which the requesting employee may have performed work.

Records that contain physical specimens, such as blood and urine samples, may not be considered as medical records, along with records concerning health insurance claims, if they are:

- Maintained separately from your medical program and its records
- Not accessible by employee name or other personal identifiable information

The industrial hygienist serves as the company's representative between the medical community, the company, and the worker. As such, it is the responsibility of the industrial hygienist to facilitate the storage and maintenance of medical records, as required by the applicable regulations, and those associated with the company or institutional process.

7.9 CASE STUDY 1: CHROMIUM IV EXPOSURE

As an industrial hygienist, you are employed by an electroplating manufacturer that uses heavy metals in the production of tools. Several workers have approached you at different times over the past year complaining about coughing and rashes that have appeared on their skin. At first, you believe there are some people with allergies associated with the different products that are used in the manufacturing business. As a industrial hygienist, you have observed work being performed and have not seen any tasks that would cause a potential overexposure. You did observe one employee not wearing gloves and quickly told the worker to leave the area and don their personal protective equipment, but you had not observed workers not wearing the required personal protective equipment per the normal work practice. In the past, the work crew were always wearing their respirators whenever you observed their work.

One employee has just returned from their annual medical physical and brought medical paperwork stating that he has developed an allergic reaction on the skin – most likely caused from coming into contact with some type of metallic substance. Based on the medical information, you interview the employee to review the work tasks that have been performed since their last physical and inquire as to what he believes is causing the irritation. The employee states that the company recently introduced a new product into its manufacturing line, but they were not sure what chemical constituents were in the product. Based on feedback from the employee, you review information associated with the employee's job duties with the supervisor.

In addition, you review air sample results that have been taken over the past six months in the general area where the employee works, along with the results of any personal air sampling that had been performed on workers in the same SEG. You also ask the employee to describe how they performed the job tasks. After reviewing the air sample results, which appear normal, you also observe how the worker performs their job and review product literature and chemical material safety data sheets to gain an understanding of what the new chemicals or products could have introduced into the manufacturing process. You confer with the occupational health medical provider and request additional testing, including targeted blood tests.

Based on the results of all the tests and information, both the occupational health medical provider and the industrial hygienist realize that the worker has been inadvertently overexposed to chromium IV. You then request a review of the manufacturing process being used to develop the new tool, and further evaluate the type of personal protective equipment being used, along with work restrictions and additional medical testing of other employees who are in the same SEG. Further testing by the occupational health medical provider resulted in identifying two additional workers who also may have been overexposed to chromium IV.

7.10 CASE STUDY 2: BERYLLIUM EXPOSURE

As an industrial hygienist, you are employed in the chemical manufacturing business. The company you work for produces chemicals used in the fertilizer business, as well as manages chemical storage facilities. You have been working for the company for more than six years and have a good understanding of all the hazards associated with the business, to include the chemicals that are used as part of the manufacturing process. A worker comes to you with a concern regarding physical symptoms and difficulty breathing and has developed pneumonia that has been difficult to treat. The worker asks for an evaluation of whether he may have been overexposed to some chemical that could be causing the pneumonia or some other problem in their lungs.

The industrial hygienist reviews the work history of the employee and discovers that the worker also performs maintenance on several tanks used to store chemical waste, along with general maintenance of the manufacturing facilities. You speak to the supervisor to find out what tasks the worker has been assigned and whether the supervisor has noticed anything unusual when assigning personnel. A review of the air sample data in the worker's general work area does not reveal anything unusual. You also review the equipment the worker uses to perform their job and discover that he recently found some old tools and started using them. The tools are non-sparking ones because of the chemicals used in the manufacturing process. Research into the tools shows that they had been stored in the back of the maintenance building for 30 years. As a precaution, you take a surface sample of the tools and ask the occupational health medical provider to perform additional blood tests to determine whether the employee could have been exposed to some type of metallic contaminant. Results of the medical testing determine that the employee has developed a sensitization to beryllium. In addition, surface sampling of the tools identified extremely high levels of beryllium. As a result of the tests, you remove the contaminated tools from the workplace, place the worker on restriction from any other exposure to beryllium, and work with the occupational health medical provider on how to best treat a short-term overexposure to beryllium and invoke a longer-term medical surveillance program, related to management of workers potentially overexposed to beryllium. In addition, you communicate the situation to the rest of employees and emphasize that a worker may not realize that overexposures to chemicals can occur not just from the manufacturing process itself, but also from unintentionally working with material that was not thoroughly evaluated in the hazard identification and control process.

Questions to Ponder for Learning

- Identify and list some of the reasons why an industrial hygienist is a key player in implementing a medical monitoring and surveillance program.
- Describe the relationship between the industrial hygienist and the occupational health medical provider.
- Identify and discuss the elements of a medical monitoring and surveillance program.
- Discuss the rights of employees to have access to their medical records.
- Review and discuss the actions of the industrial hygienist performed in investigating the potential exposure that occurred in Case Study 1.
- Review and discuss the actions of the industrial hygienist performed in investigating the potential exposure that occurred in Case Study 2.
- What additional actions should the industrial hygienist have taken to identify and quantify the potential health risks in Case Study 1?
- What additional actions should the industrial hygienist have taken to identify and quantify the potential health risks in Case Study 2?
- What are the health-related impacts of chromium IV and beryllium, and what controls could have been identified to prevent the possible overexposures?

REFERENCE

1. Part 10 Code of Federal Regulations Subpart 851, Appendix A, Section 8, *Occupational Medicine.*

8 Workforce Training on Hazard Recognition and Control

8.1 INTRODUCTION

As recognized by the ISO 4500:2018,[1] training can be an effective way to ensure that employees are integrated into the workplace and be a significant component in the overall risk reduction process. Providing training to workers will acclimate them to workplace culture, policies, and procedures, and improve their ability to work safely, which is the responsibility of the employer. There are times when employees will receive training once on a hazard, and there are other times when training is required at some frequency, for example, annually or each time a worker is reassigned to a new job. It is the responsibility of the employer to ensure that an adequate hazard recognition and mitigation training program is in place and that employees are trained. Training is a condition of employment for most employees, and refusal to participate in learning activities and initiatives is not optional and can result in termination of employment.

Training is crucial for organizational development and productivity. It is beneficial to both employers and employees, regardless of the market or profession in which they operate and compete. An employee will become more efficient and productive if they are knowledgeable of the task that they are performing. This knowledge also translates into a worker's ability to make decisions to work safely and look out for the safety of coworkers.

The industrial hygienist's primary focus is on eliminating or reducing workplace hazards and the reduction of risk to physical, chemical, and biological stressors found in the workplace. These professionals typically gain extensive training in ways to make hazard recognition and mitigation decisions while completing their college education and through internships within the industry.

Providing the necessary training overall creates a knowledgeable staff of employees that can provide the flexibility to relieve one another as needed to work on teams or independently. Training can also build an employee's confidence because he or she has a stronger and perhaps broader understanding of the company and their job responsibilities. This confidence can put a worker in a position to excel in performance while remaining engaged in the business.

Employees can become injured if they are focused only on following the rules that have been put in place and not on the actions in implementing those rules. Employees

DOI: 10.1201/9781032645902-8

have a better chance of working safely and injury-free if they possess the skills and knowledge needed to discern or anticipate and recognize workplace hazards and associated health risks. Providing hazard recognition training to all employees is the best way to improve safety, and it personalizes safety for the workers to the point that they will become engaged and willing participants in their safety, and the safety of their coworkers. Despite the up-front and ongoing investment, training and development provide both the company, and the individual employees, with benefits that make the cost and time a valuable investment.

8.2 WHY PROVIDE WORKPLACE TRAINING?

Training is generally used as an opportunity to refresh and expand the knowledge base of employees across all business sectors. However, many employers are of the opinion that training development opportunities are expensive primarily because employees often miss out on work time while attending training, which can cause delays in the completion of work activities, as well as the associated cost of training delivery. Per ISO 4500:2018,[1] Section 7.2, *Competence*, the organization is required to:

- Determine the necessary competence of workers that affects or can affect its occupational health and safety (OH&S) performance, which includes defining the level of competency required to safely perform the work.
- Ensure that workers are competent (including the ability to identify hazards) on the basis of appropriate education, training, or experience.
- Take actions, where applicable, to acquire and maintain the necessary competence, and evaluate the effectiveness of the actions taken.
- Retain appropriate documented information as evidence of competence.

A proactive company or organization will view training as a key function because it reduces financial and professional risks and liabilities and can improve productivity.

There are many workplace hazards that require worker training to ensure that they are aware of the hazards involved, the associated health risks, and how to safely perform work. Table 8.1 provides a general list of well-recognized hazards in the workplace. In the United States, most of these hazards have specific hazard communication and training requirements stated by the Occupational Safety and Health Administration (OSHA) and international safety regulatory agencies. Training is important in addressing hazards such as the ones listed in Table 8.1.

Training is also a key component in addressing the following:

- Improvement of worker performance and productivity: An employee who receives the necessary training has confidence in their abilities and is better able to make decisions and perform their job responsibilities. Employees have the knowledge and skills to help them perform their work safely, reduce their exposure to health risks in the workplace, and properly implement policies and procedures.

TABLE 8.1
Example of Well-Known Industrial Hazards

Asbestos	Blood-borne pathogens	Electrical	Fire
Noise	Hazardous chemicals	Confined spaces	Slips, trips, and falls
Ionizing radiation	Nonionizing radiation	Lead	Carcinogens
Compressed gases	Vibrations	Trenching	Solvents
Chemicals	Chemical solutions	Carbon dioxide	Carbon monoxide
Welding	Working from heights	Illumination	Hexavalent chromium

- Weaknesses in knowledge and skills: An effective training program supports the strengthening of skills that employees need to improve and perform their jobs efficiently with the required quality. A training program designed to develop workers will move employees to a higher level of performance by providing them with similar skills and knowledge.
- Improvement and facilitation of consistency: A well-structured training and development program ensures that employees have consistent experience and knowledge in a specific topic or at a required skill level.
- Improvement of employee satisfaction and retention: Employees with access to training and development opportunities have an advantage over employees in other companies who are forced to seek out and fund training opportunities on their own. The investment in training demonstrates to employees that they are valued and is helpful in gaining support from workers, and at times retaining them as engaged employers.
- Reduction of accidents and the health risks associated with the performance of work. Training reduces human errors, which can potentially occur when employees lack the knowledge and skills required for completing a particular task. When workers are appropriately trained, the chances of being involved in an on-the-job accident decrease and the employees will become more proficient in performing their job.

8.3 DEVELOPING AN EFFECTIVE TRAINING STRATEGY

An effective training strategy is vital for companies to be able to impart knowledge to the workforce and ensure that the quality of their products meets company expectations for performance. Developing and implementing a strategy for training can provide a competitive advantage for a company. The training plan needs to be comprehensive and have the support of management at all levels. Critical steps to consider when developing a training strategy include:

- Meeting with the leadership team to ensure that the training strategy being developed complements the company business strategy and supports the mission statement.

- Ensuring that the training strategy is developed to identify and address training gaps, as well as being supportive of new training needed to develop knowledge and understanding of new policies, processes, or procedures.
- Establishing learning objectives that will provide the desired training outcome.
- Identifying and deploying a learning management system that will complement the administration and tracking of training programs, to include course attendance and completion.
- Once the strategy has been developed, reviewing the plan with the leadership team and discussing it with some workers to gain support its implementation.

During development of the strategy, it is pertinent that the leadership team be engaged in the development of training material so that buy-in can be achieved, funded, and the strategy can be implemented with ease. A training strategy lacking leadership buy-in will not be effective and will not serve the needs of the company. It would also be helpful if input from the workers is solicited since the workers will be involved in the training program through course attendance.

8.4 HAZARD RECOGNITION, EVALUATION, AND CONTROL TRAINING

Training employees on workplace hazards is the first step in protecting workers, ensuring compliance with regulatory requirements, and ultimately reducing health risks to the workers. In such a case, knowledge is truly a powerful tool to ensure worker safety. In fact, the OSHA, and the ISO 45001:2018, and organizations mandate that employees be made aware of workplace hazards, as well as the necessary means to protect themselves from those hazards.

To demonstrate compliance and aid in risk mitigation, employees must be competent enough to participate in the performance of four key risk management steps designed to enhance the feasibility of performing work safely. These steps are shown in Figure 8.1 and are further defined in Sections 8.4.1–8.4.4. An effective training program is designed to enhance employees' ability to successfully navigate through these steps and participate in the planning and performance of work safely. Understanding these key steps is essential in reducing and controlling risk for workers while performing their daily tasks, as well as risk to the company.

One of the best ways to develop and implement a training course geared toward dealing effectively with the four key risk management steps is to develop training that is based on evaluating an actual task or process that the workers are expected to perform, as well as including those that may not be applicable to their daily tasks. Walking through real-life processes or cases with trainees will guide them through the thought process and is likely to help them cultivate ways of anticipating issues before they arise, recognize hazards when they arise, evaluate hazards, and develop creative and effective solutions to mitigate hazards.

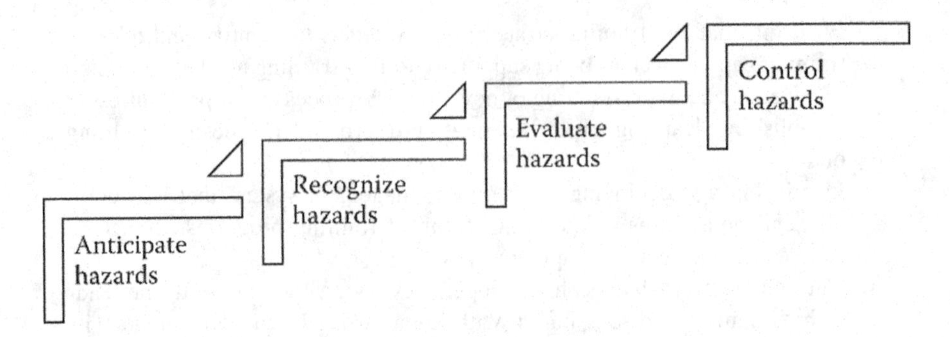

FIGURE 8.1 Key risk management steps.

8.4.1 Hazard Anticipation

The ability to anticipate hazards requires knowledge of the process or task, as well as the ability to draw on previous experiences and perhaps the professional judgment of the subject matter experts (SMEs) and workers involved in the process. Developing and implementing training that will assist employees in their ability to anticipate hazards involves drawing on their knowledge and skills, and their ability to see beyond what is obvious. Workers must be trained on ways to be inquisitive and look at what can potentially happen even if the task has been done the same way many times without incident. Inquisitiveness is important for the following reasons:

- It activates the mind. Inquisitive people tend to ask questions and search for answers, keeping their minds active.
- It opens the mind and allows the mind to be observant to new ideas. The mind begins to expect and anticipate new ideas.
- It opens new possibilities and ways of doing things.

Since it is difficult to train one on how to demonstrate inquisitiveness, it is pertinent that the training includes a discussion of the tools that can enhance employee inquisitiveness and hazard anticipation. Hazard anticipation is most effective when everyone in the business environment is actively engaged in seeking out situations that can pose a hazard to workers. Table 8.2 lists some roles that a team can play in anticipating hazards for optimum results.

Below are ways to enhance one's ability to become and remain inquisitive so that hazard anticipation can take place:

- Ask questions. Asking questions creates opportunities for discovering new information. Questions that may be beneficial include:
  What can go wrong?
  Does this procedure optimize the performance of the task safely?
- What can happen during the performance of the task that may create a hazard?

TABLE 8.2
Support Role in Hazard Anticipation

Support Role	Contribution
Employees (SMEs, engineers, project managers, etc.)	Employees that have knowledge and experience in performing a task is key in helping anticipate what can potentially go wrong, based on their knowledge and hands-on experience
Management	Management is key in ensuring that workers have the appropriate training and skills to enable them to seek out hazards to prevent injury. Management must also set the stage and expectations for workers to freely identify hazards, with the expectations that management will respond positively. Employees must feel free to disclose the possibility of issues occurring before they occur to management; otherwise, some hazards that may arise will not be anticipated and dealt with in advance of occurring
Industrial hygiene professional	These professionals, by nature, have been trained to anticipate hazards using inquisitiveness, knowledge gained through education, and professional judgment. They have been specifically trained in the hazard recognition and control process and risk mitigation techniques

- Can this task be performed differently to reduce risk to workers?
- Listen openly and without judgment. This allows for receptivity of conversation and consumption of new information.
- Expect and be willing to embrace the unexpected.
- Be willing to let go of the old way of doing things if a newly discovered way provides better results.
- Be willing to expect and embrace change.

8.4.2 Hazard Recognition and Identification

As one would expect, the most effective and comprehensive way to be able to identify hazards is having participants with experience and knowledge in the industry and the projects performed by the company. The training developed to assist workers in their ability to identify hazards in the workplace should take advantage of the knowledge of the industry that the workers may already have. However, do not assume that they have all the knowledge needed to recognize hazards just because they are skilled in preforming their daily work responsibilities. Hazard recognition training goals should include:

- Providing an understanding of the project goals, processes, and the tasks that are being performed.
- The types of hazards known in the workplace.
- Knowledge of the various regulatory requirements outlining knowledge requirements to perform work safely, to include the safe use of equipment.

- The knowledge necessary for workers to be able to break work down into distinct tasks, so that each task can be evaluate and hazards identified.

Hazard recognition training should address all categories of hazards, such as those listed below:

- Chemical: Gas, liquids, and vapors.
- Physical: Mechanical, electrical, and work area (slips, trips, and falls).
- Biological: Mold, bacteria, and viruses.
- Ergonomic: Lighting, repetitive motion, and awkward body positioning.
- Other: Workload, distractions, conflicts, tools, use of personal protective equipment, and cold and hot environments.

Table 8.3 lists the roles that a team can play in recognizing workplace hazards for optimum results.

8.4.3 Evaluate Hazards

Successful hazard evaluation methods require the knowledge and input of skilled participants involved in planning and performing the work. It is a team effort that renders the process successful and allows for diversity of thoughts and opinions that can lead to an appropriate mitigation solution that will be successful. Hazard evaluation is at the core of every work planning process; it is a process that workers must be trained in, as well as having the capability to participate in during the initial planning stage. Too often, workers are left out of the planning process and are expected to complete work based on what others believe to be the right approach. Leaving

TABLE 8.3
Support Role in Hazard Recognition

Support Role	Contribution
Employees (SMEs, engineers, project managers, etc.)	Employees are generally the best set of eyes in recognizing hazards, since they are up close with the work, day in and day out. These individuals can add specific information gained from historical knowledge and experience that may not be easily detected during the hazard analysis process
Management	Management is responsible for ensuring that workers receive the appropriate training to enhance their safety. Training that provides employees with the skills and knowledge needed to assist with hazard recognition is a key attribute of keeping them safe while performing work
Industrial hygiene professional	Hazard recognition is a large component of the job responsibility for industrial hygiene professionals. The training they receive is geared toward enhancing their ability to aid in the process. Their contribution is key to ensuring that the process is completed in totality with the highest level of accuracy possible

TABLE 8.4
Support Role in Hazard Evaluation

Support Role	Contribution
Employees (SMEs, engineers, project managers, etc.)	Employees are engaged in frequently making decisions on hazard evaluation as they are performing their jobs. They are faced with making real-time decisions on whether to continue a task if an unsafe condition is introduced or discovered. Using their knowledge, employees make decisions on whether to continue a task when changes arise or pause work and contact management
Management	Management support is necessary in informing employees that their expectation is that hazards are evaluated, and workers are important to the process, and that they should alert management when they feel that the process is not being performed or is not performed to the level that will facilitate safe performance of work. Management secures funding for industrial hygiene and worker training
Industrial hygiene professional	Hazard evaluation is a key role that industrial hygienists are trained to perform. The industrial hygienist should have gained training during the process to become an industrial hygiene professional. However, there may be times when these professionals will require additional training or briefings on the operability of a process or task that will provide the knowledge needed to add value to the hazard mitigation process

workers out of the planning process is an accident waiting to happen. Table 8.4 shows the importance and contribution of various roles in the hazard evaluation process.

Training should be comprehensive and provide the skills for workers to effectively use work breakdown concepts to be able to identify all tasks and the associated hazards. In addition, workers must be able to identify when performing a task can impact the hazard profile of another.

8.4.4 Controlling Hazards

Once a hazard has been identified and understood, hazard mitigation is the next step which is not always easy and may be costly. Prior to placing workers in harm's way, mitigation is a necessity. Hazard mitigation oftentimes requires the insight and knowledge of a team that most likely includes management, the worker, and the safety and health professional. Hazard mitigation training must be industry specific, company specific, and task specific in order to be effective. It is not always feasible to completely remove the existence of a hazard during the performance of a task; therefore, it must be recognized during the hazard/risk evaluation process when the use of personal protective equipment and administrative measures are required to control the risk of a hazard to the worker. Table 8.5 shows the contribution of various roles in the hazard mitigation process.

As defined in ISO 45001:2018, Section 8.1.2, *Eliminating Hazards and Reducing OH&S Risks*,[1] efforts to mitigate hazards can include the following:

TABLE 8.5
Support Role in Hazard Mitigation

Support Role	Contribution
Employees (SMEs, engineers, project managers, etc.)	Employees can provide insight into means to mitigate hazards, as they are closer to activities and oftentimes have improvement measures that are already implemented at some level informally
Management	The leadership team with specific knowledge on how work is performed can add value to the mitigation process. They must be onboard with the solution, as many times the solution will impact productivity or require additional funding to implement. Also, they must be willing to pay for additional training for the industrial hygiene professional
Industrial hygiene professional	Effective industrial hygiene professionals obtain knowledge through their training that will aid them in successfully working with employees and management in mitigating hazards. However, there may be times when these professionals will require additional training or briefings on the operability of a process or task that will provide the knowledge needed to add value to the hazard mitigation process

- Hazard elimination
- Substitution of chemicals
- Installation of engineering controls
- Process redesign
- Changing the sequence of steps when performing a task (administrative controls)
- Use of a respirator (respiratory protection as the last level of defense to be used in protecting workers from workplace hazards)

During the hazard recognition and mitigation process, it is important for all hands and minds be on deck. This means that employees, supervisors, and the industrial hygienist must be actively engaged in the process. Active engagement includes exhibiting the appropriate knowledge level, which enables participants to provide meaningful input.

8.5 TRAINER KNOWLEDGE AND QUALIFICATION

It is important that the person delivering the training be knowledgeable of the topic and have credibility to be viewed as an effective trainer. It is not feasible to select a person with no topical knowledge or experience and expect him or her to be viewed as being credible. Trainers must have some knowledge of the hazard recognition, evaluation, and mitigation processes. They should also be knowledgeable of techniques that can be used to achieve results.

Trainers delivering training without knowledge of the subject are not likely to be effective. What makes a trainer effective?

- Preparation
- Knowledge and experience in the topic being taught
- Clear communication skills
- Ability to connect with the students
- Ability to capture and retain the attention of students
- Ability to make the training relevant to the trainees' job and life outside of work
- Ability to engage students in the topic
- Ability to introduce stories that will complement the training and connect workers to the course topics

8.6 TRAINING EFFECTIVENESS EVALUATION

There must be a process in place to ensure that the training that is being delivered to the workers and supervisors is effective and meets the need of the business. Training should also be evaluated at some frequency to ensure that the strategy and field execution of a task is in alignment with the training being offered. The most popular means used to evaluate training is Donald L. Kirkpatrick's method for evaluating training programs.[2] This model focuses on four aspects of training: reaction, learning, behaviors, and results. Each of these stages is briefly described in Table 8.6. Whatever model or steps are used to evaluate the effectiveness of training, evaluating training quality is a must to ensure that objectives are being met.

TABLE 8.6
Donald L. Kirkpatrick's Four Steps for Evaluating Training Effectiveness

Stage	Description
Reaction	The result of this measurement sheds light on how the trainees feel about the course. Do they believe that the course achieved its objectives? Have they benefited from the course in terms of completing their jobs more safely?
Learning	Determining whether the material delivered was absorbed and understood by the trainee. The important question to answer is whether the training delivered knowledge and this was understood by the trainees at the time
Behaviors	Recognizing that it is not always easy to measure a change in behavior. However, it is beneficial to measure or determine whether the training is helpful in moving attendees to modify their behavior in a positive way
Results	Important aspects to include in the measurement of results include improvement in productivity and quality, reduction in accidents and injuries, and employee engagement

8.7 OTHER TRAINING METHODS AND TOOLS

There are other methods that can be used effectively as training tools that may not be widely thought of and used. These methods include:

- Process or tool mock-up
- Peer-to-peer or on-the-job training
- Training at the job site
- Hazard identification checklists
- Case studies and storytelling

Each of these tools has distinct benefits, as well as challenges. Therefore, choosing to use one over the other must be done with careful consideration.

8.7.1 Process or Tool Mock-Up

Mock-up training is generally instituted when the task to be performed is the first of its kind or the hazards are so great that it is incredibly important that all involved become extremely knowledgeable in how to carry out the task safely. A great example of mock-up training is establishing a mock-up operational control panel, similar to the actual control room, that can be used to practice responses to different operational plant conditions or emergencies. Another example of mock-up training is establishing an area for IH professionals to practice air sampling in a confined space. This method can be expensive and time-consuming because the exact conditions must be created, and the task performed in the same manner during the mock-up as it will be during the actual performance. This method is effective because it:

- Provides hands-on training for the worker
- Gives the worker the ability to learn the entire task in totality, perform the task in a simulated environment, and become proficient without being exposed to the hazards that are involved in the actual task
- Provides the worker with needed confidence to complete work
- Improves the worker's ability to complete work safely through practice

8.7.2 Peer-to-Peer Training

Peer-to-peer training is the process of training delivered to workers by their peers, with the peers having had the opportunity to perform the tasks that the new worker is expected to perform. These workers serve as a key contributor in assisting new employees in becoming acclimated with the hazard recognition and mitigation process through their experience and knowledge of the work that is being performed, as well as the expectation of the leadership team. Careful attention must be paid to ensure that the training provided is in accordance with procedures, because if a task is performed inadequately by a peer trainer, the new worker will be improperly trained and will in turn incorrectly perform the task which can present a hazard to the worker. A great example of peer-to-peer training is the HAMMER Facility at the U.S. Department of Energy Hanford Site. The HAMMER Complex was established in 1997 and provides realistic, hands-on, up-to-date training to Hanford Site workers, military, national, and international emergency responders, and Homeland Security personnel. The HAMMER Complex is situated on an 88-acre campus and uses both craft and exempt professional staff to train personnel. Many of the training classes are taught by craft personnel who may be temporarily or permanently assigned to support training in their established discipline.

8.7.3 Training on the Job Site

Training at the job site involves dispatching a qualified trainer to the job site to train workers on processes and tasks in the environment in which work is to be performed. The advantages of conducting training in the work environment include:

- The opportunity to evaluate hazards associated with the work, as well as any hazards that may be introduced by the work environment.
- The opportunity to witness firsthand the types of distractions that may affect safe performance of work.

Disadvantages of this method include:

- Disruption of work going on in the work area
- Distraction to workers
- The potential creation of a hazard by having other untrained workers in the work area
- Loss of productivity

8.7.4 Hazard Identification Checklist

Hazard identification checklists are widely used by many industries and businesses to identify workplace hazards. These checklists have been, and continue to be, used as a means to train and engage workers in the hazard identification and resolution process. Hazard identification checklists are a great way to train workers to be alert and focus on activities or situations that can increase risk to the worker. These checklists can be written simplistically, in a way in which they are easily understood and used. An example of a hazard identification checklist is shown in Table 8.7.

8.7.5 The Use of Case Studies to Enforce Learning

Earlier in this chapter, it was stated that a good way to deliver training is to use case studies and communicate real life stories to connect the trainee with the topic. Case studies when presented appropriately can be used as a form of storytelling and increase the effectiveness and the retention of material being introduced to learners. Storytelling is gaining popularity in delivering training in the corporate environment

TABLE 8.7
Hazard Identification Checklist

	Yes:	No:	Date corrected:
Compressed Gas Cylinders			
1. Stored in racks or carts			
2. Tanks/cylinders inspected			
3. Clearly labeled as to the content			
4. Stored with the protective cap in place			
5. Secured in place to avoid tipping or falling			
6. Stored at least 20 ft away from flammables			
7. Cylinder valves closed when not in use and before moving			
Hazardous Materials			
1. The inventory list and SDSs are available			
2. All hazardous materials are properly labeled if transferred from the original container			
3. Containers are in good condition			
4. Employees were trained on the hazards of the chemical substances			
5. Hazardous substances are used in well-ventilated areas			
6. Appropriate personal protective equipment is available to be used when handling chemicals			
7. Flammable chemical containers are stored in a flammable cabinet			
8. Flammable containers are closed when not in use			

because of its effectiveness. Stories have the capability of capturing the attention of learners and stimulate their creativity as well as appealing to their emotions. The case study below presents a viable example to demonstrate the role of case studies or discussions of actual tasks and may be useful in the hazard training process.

8.7.5.1 Case Study

Company AA has embarked on a project to install a gas pipeline to support the increasing population in a large urban community. The project is being completed and discussed in eight steps representing eight distinctive tasks that must be evaluated and hazards that must be mitigated. Table 8.8 presents an example of the thought process used to analyze the hazards associated with the case study work steps in Table 8.9.

8.7.6 Microlearning

The microlearning philosophy has been gaining traction and popularity as knowledge of the concept is shared by companies and educators the benefits it can bring

TABLE 8.8
Example of Hazard Anticipation, Recognition, Evaluation, and Mitigation Thought Process

Work Package Step Number	Anticipate Hazards	Recognize Hazards	Evaluate Hazards	Mitigate Hazards
1	Will there potentially exist poisonous insects and plants?	Poisonous plants and insects seen in the pipe installation path	What type of poisonous plants and insects are in the area?	Keep skin covered to avoid contact with plants. Heightened awareness for poisonous insects
1	Will heavy equipment be used in the same area where people will work?	Potential for human–machine contact when workers are in the work area when heavy equipment is used	Can workers and heavy equipment be isolated?	Barricade the heavy equipment when in use to keep workers from coming in contact with it
1	Will weather and environmental conditions pose hazards?	Potential for adverse weather conditions when working outside on the project	What type of weather hazards, and associated industrial hazards, can we expect this time of year or season?	Train personnel on how to protect themselves from hazards associated with weather and ensure each worker has the tools needed to protect themselves
2	Can a worker become injured by coming into contact with the trailer?	Unstable load on trailer	At what point during the task will workers come into contact with the trailer and load, representing a hazard?	Ensure that the load is strapped down when materials are stored on the trailer
2	Is there potential for the employee to come into contact with the trailer and load?	Material handling	What activities, including material handling, could cause injury and how could we prevent the injury?	Use proper lifting techniques when handling material. Use mechanical lifting for material weighing more than 50 lb

(*Continued*)

TABLE 8.8 (CONTINUED)
Example of Hazard Anticipation, Recognition, Evaluation, and Mitigation Thought Process

Work Package Step Number	Anticipate Hazards	Recognize Hazards	Evaluate Hazards	Mitigate Hazards
3	Will workers have the need to lift pipes?	Lifting/material hazard	What work activities requiring manual lifting?	Use mechanical lifting for material weighing more than 50 lb
3	Can equipment and workers come into contact during the task?	Employee contact with excavator. Open surface/ trench. Noise	Noise level above 85 decibels. Training required to enter the trench. Employees work within the proximity of the equipment	Use mechanical lifting for material weighting more than 50 lb. Provide training on trenching for workers entering the trench. Barricade the excavator when in operation. Wear prescribed PPE: gloves and hearing protection device
4	Will worker contact with the hydraulic bending machine create a health hazard?	Worker hands will come in contact with machine. Debris coming in contact with the eyes. Pipe moves during bending	Pinch point. Eye injury. Noise. Striking or crushing injuries	Wear personal protective equipment, heavy work gloves, eye protection, a hard hat, safety shoes, and hearing protection. Avoid placing hands in between the pipe and machine. Ensure adequate clearance around the machine when operating

(*Continued*)

TABLE 8.8 (CONTINUED)
Example of Hazard Anticipation, Recognition, Evaluation, and Mitigation Thought Process

Work Package Step Number	Anticipate Hazards	Recognize Hazards	Evaluate Hazards	Mitigate Hazards
5	During performance of the job, will workers be exposed to radiation?	Radiation from x-raying pipes	X-ray burns	Ensure workers are trained to work with radioactivity. Apply radiological control practices to reduce and monitor exposure. Maintain positive access control when performing x-ray process
5	Will employees be exposed to hazards associated with welding?	Welding process hazards. Uneven surfaces	Arc flash from welding. Slips, trips, and falls	Ensure that workers are not in the work area during welding. Employee training. Welder use of engineering control (local ventilation) and PPE (leather gloves, welder's shield, and apron)
6	Can the coating present a hazard to workers?	Potential skin irritation. Eye contact	Skin irritation from contact with coating. Eye contact from splash	If feasible, select no hazardous coating. Wear gloves and long sleeves to avoid skin contact with chemical products
6	Will lifting of pipes be involved, and can workers potentially be hurt as a result?	Pipe handling	Lifting hazards from handling pipes	Use mechanical means when lifting more than 50 lb
7	Is there any opportunity for the tractor to come into contact with the worker during operation?	Employee contact with tractor. Noise	Employee contact with tractor during operation. Noise level above 85 decibels	Barricade area around equipment when in use. Wear hearing protection. Verify operator training

(*Continued*)

TABLE 8.8 (CONTINUED)
Example of Hazard Anticipation, Recognition, Evaluation, and Mitigation Thought Process

Work Package Step Number	Anticipate Hazards	Recognize Hazards	Evaluate Hazards	Mitigate Hazards
7	Can a worker become injured by coming into contact with the front-end loader?	Employee contact with the end loader during operation. Noise. Dust	How and when will the employee come into contact with equipment during operation. Noise level above 85 dBA. Control dust generated from excavation	Barricade area around equipment when in use. Verify operator training. Wear hearing protection. Use water to keep soil moist to control dust
7	Will there be power lines in the area?	Power lines are located above ground up the road from the excavation	Locate of above ground power lines and identify in work steps	Avoid contact with power lines and maintain 50 ft distance from lines
8	What hazards are associated with backfilling using a front-end loader?	Dust hazard to eyes, nose, and skin. Personnel could be in contact with the equipment	Dust. Contact hazards with equipment	Use personal protective equipment and wear long sleeves. Establish barriers (access control) while the equipment is being used

Note: This table does not contain all the hazards that may be present during the performance of the tasks involved in completing a project. It is used only as an illustration of the process. PPE, personal protective equipment.

to the learning process. Microlearning, sometimes referred to as bitesize learning, refers to a type of training that is designed to enhance comprehension and retention of learning material. The demand for development of training through microlearning has significantly increased due to the ability for employees, or people, to increase their ability to learn through smaller, focused learning modules that are focused on one learning objective or topic. These training modules or segments are designed to be less than 20 minutes and are typically available on demand, such as apps that teach a foreign language, via an electronic table, computer, or phone.

TABLE 8.9
Case Study Work Steps

Step 1: Site preparation. Trees and shrubbery must be removed, the ground must be leveled, and the surface soil must be separated from the subsoil. The subsoil will be returned after the area has been backfilled to support healthy plant growth.
Step 2: Trailers are used to transport pipes and supplies, and a special excavator is used at the work site.
Step 3: An excavator is used to dig the ditch where pipes will be placed. The depth and width of the ditch are solely dependent on the pipe size and the project specifications.
Step 4: Pipe bending will take place to accommodate curved locations and ditch elevations to conform to topography. This task is performed using a specialized machine designed to bend pipes with ease.
Step 5: Pipe welding and x-ray. During the project, pipes are welded to join them together into one continuous length. To ensure integrity, each weld is x-rayed.
Step 6: Coating of pipes to prevent rust. Pipes are coated at the job site using an epoxy and high-density polyethylene coating.
Step 7: Tractors are used to lower the pipes into the trench.
Step 8: Backfilling is carried out using a front-end loader.

Training through microlearning when electronically delivered allows a person to train at their convenience, increasing flexibility in when the training can be performed. When developing microlearning for IH topics, the learning modules must be fundamental. One objective per module, the training material cannot be complex. Graphics that are engaging will enhance the user experience and ability to retain the training material.

Questions to Ponder for Learning

1. Describe the importance of an effective training program in facilitating a safe work environment.
2. Discuss some workplace deficiencies that training can be used to address.
3. What key steps are important when developing a training strategy?
4. Discuss the effectiveness and value of an inquisitive workforce when hazard recognition, control, and mitigation are goals.
5. List and discuss goals of an effective hazard recognition training course.
6. What roles does management play in the hazard mitigation process and training?
7. List and discuss the attributes of an effective trainer.
8. What is a potential issue that may arise as a result of peer-to-peer training?
9. What is microlearning and what is an example of microlearning when discussing the hazard recognition process?

REFERENCES

1. International Standard, ISO 45001, 2018. *Occupational Health and Safety Management Systems – Requirements with Guidance for Use*, International Organization for Standardization.
2. Fitzpatrick DL, Kirkpatrick JD, 2006. *Evaluating Training Programs: The Four Levels*, 3rd Edition, Berrett-Koehler Publisher.

9 Industrial Hygiene and Emergency Response

9.1 INTRODUCTION

The role of the industrial hygienist in responding to an emergency is gaining importance, both domestically and internationally, as emergencies are becoming more common at work and in our private lives, within the Unites States and throughout the world. All one has to do is pick up a newspaper or connect to the internet to understand the types of threats the United States and the international community are facing. Historically, fire and medical departments have been considered the response organizations relied on when a disaster or accident occurs. They are still generally the first responders, but as the use and reliance on technology and communication/information exchange has improved and increased over time, the roles of first responders have also become more specialized to improve the quality of the response provided, including the use of industrial hygienists when responding to chemical or safety emergencies. Below is a summary of several emergency events that have occurred over the past 20 years.

9.1.1 Hurricane Katrina – United States

Hurricane Katrina impacted the Gulf Coast of the United States on August 29, 2005 and involved a very significant emergency response effort. At the time of striking the coast, the hurricane had been designated a Category 3, which means that devastating damage can occur (the worst rating is 5). Although the hurricane ranking system provided an indication as to the strength of the hurricane, what was not well understood was that the hurricane rating scale did not address the potential for other hurricane-related impacts (secondary impacts), such as storm surge, rainfall-induced floods, and tornadoes. In the case of Hurricane Katrina, the storm itself caused significant damage, but the aftermath of the storm was what really drove the massive damage – which is still being felt today. A large part of the damage was caused by rain and flooding that occurred before and after the hurricane. In particular, the heavy rains overwhelmed levees and drainage canals in New Orleans and water flooded the city. Emergency response to Katrina included industrial hygienists who performed exposure monitoring and identified preventative hazard controls to decrease the spread of unhealthy air and water. Estimates vary, but a significant number of people died as a result of this disaster, and many thousands were displaced from their homes and continue to rebuild their lives.

DOI: 10.1201/9781032645902-9

9.1.2 La Porte, Texas Chemical Plant – United States

On November 15, 2014, an explosion occurred at a chemical plant in La Porte, Texas, where insecticides were produced. A few days before, an inadvertent chemical dilution caused operating difficulties forcing a shutdown of DuPont's Lannate® Unit. In response, personnel adjusted the unit's control system to resume operations. On November 12, 2014, operators tried to restart the equipment; unfortunately, methyl mercaptan piping became plugged. At that time, the operators did not know the source of the plugging and began troubleshooting. During troubleshooting, methyl mercaptan was released both outside and inside the manufacturing building, triggering 32 methyl mercaptan gas alarms on the control panel throughout the 17 hours preceding the incident. They did not perceive the methyl mercaptan alarms as signifying a serious hazard because they had normalized the methyl mercaptan odor within the Lannate® Unit, as well as the detector alarms.

Sometime between 3:01 am and 3:13 am, a worker (likely the shift supervisor) manually opened two sets of drain valves on the vent header piping, located on the third floor of the manufacturing building. The liquid methyl mercaptan that had filled the piping escaped out of the valves, vaporized, and killed the shift supervisor. In addition, the release killed three operators who died from a combination of asphyxia and acute exposure (by inhalation) to methyl mercaptan.

The cause of the highly toxic methyl mercaptan release was determined to be flawed engineering design and the lack of adequate safeguards. Contributing to the severity of the incident were numerous safety management system deficiencies, including deficiencies in formal process safety culture assessments, auditing and associated corrective actions, troubleshooting operations, management of change, safe work practices, shift communications, building ventilation design, toxic gas detection, and emergency response. Weaknesses in the DuPont La Porte safety management systems resulted from a culture at the facility that did not effectively support strong process safety performance. From the initial notification of the emergency, the event response was hampered due to miscommunication, disorganization, and a lack of situational awareness.[1]

9.1.3 September 11, 2001 – United States

One of the most significant examples of an emergency response effort in the United States was September 11, 2001 (9/11). Terrorists flew two planes into the World Trade Center (WTC) in New York City, a plane was flown into the Pentagon near Washington, DC, and an additional plane was crashed into a field in Pennsylvania. The attacks resulted in a massive loss of life and destruction of property. As part of the response and recovery actions associated with 9/11, the industrial hygienist played a key role in understanding the risks posed to response personnel and the people affected by the disaster. Recovery of the devastation caused by the 9/11 attacks will continue for years to come, along with efforts to minimize the ability for future terrorist attacks to occur, and also to improve response and recovery actions. Chapter 9.7 discusses in further detail the role an industrial hygienist played in actually responding to the event.

9.1.4 Collapse of the Rana Plaza

The Rana Plaza was located in Bangladesh and contained multiple clothing factories. Workers at the garment factory manufactured clothing and items for many successful clothing companies. The lower half of the building contained other retail stores and some housing.[2] The factory was known to have structural damage, yet people were still required to work at the building even in its unsafe condition. On April 23, 2013, personnel reported significant structural damage to the building owner and the building had been evacuated, but the owner continued to communicate the building was safe and required everyone to return to the building to work. On April 24, 2013, the building collapsed with more than 3,000 people inside the structure. As a result of the building collapse, 1,134 people were killed and more than 2,500 were injured.

Per the Bangladesh Fire Service and Civil Defense, the upper floors of the building were built without a permit and were structurally unsafe. The building's architect stated the building had been designed for retail shops and was not built to withstand the weight and vibrations of factory machinery, including vibrations from diesel generators switched on following a power outage. Many people were held responsible for the tragedy and were held accountable for the lack of diligence in monitoring safety standards and wages paid.

9.1.5 Glasgow Explosion – Scotland

On May 11, 2004, an explosion occurred at Grovepark Mills, Maryhill, Glasgow, which caused the substantial collapse of the former Mill building. A total of nine people were killed and 45 seriously injured. The premises at Grovepark Mills were owned by ICL Plastics Limited and occupied by ICL Technical Plastics Limited and Stockline Plastics Limited. Investigation into the event determined that the explosion was caused by igniting a buildup of explosive gases, liquified petroleum gas (LPG), in the building basement. The explosion caused the building to collapse within its own footprint. The Grovepark Street was severely damaged but remained standing. Several pieces of the concrete ground floor were found some distance away and blood samples from one of the deceased showed that they had inhaled propane gas before the explosion. As a result of the Glasgow event, several criminal charges were filed including[3]:

- Failing to make a suitable and sufficient assessment of the risks to the health and safety of employees while at work.
- Failing to appoint one or more competent persons to assist in carrying out such risk assessments.
- Failing to have a proper system of inspection and maintenance in respect of the LPG pipework concerned.
- Failing to ensure the pipework was maintained in a condition that was safe and without risk to employees.

The Glasgow accident reminds all of us that not only are there significant impacts to life when an industrial accident occurs, but there are also significant financial/criminal liabilities that can result.

No matter the type of accident, emergency response actions are approached in one of two ways: they will be either planned or unplanned. The distinction in the manner by which the response occurs is significant and largely drives actions associated with the coordination and preparation by the responders.

9.2 APPROACHES TO EMERGENCY RESPONSE

Planned emergency responses include the planning and preparation of identified emergency scenarios, the actions needed to mitigate an emergency, and the routine training of employees on the required response actions. Emergency response scenarios vary depending on the mission and function of the company and, in most cases, are clearly defined because of the daily operations of the company and the actions required by each person involved in response actions. In some cases, government regulations, associated with high-risk industries, require emergency scenarios and response actions to be well defined, rehearsed, evaluated, and successfully implemented.

During a planned emergency response, actions such as the number of resources needed to successfully mitigate the hazard caused by the emergency have been identified, and equipment needed to assist responders is stored in a ready accessible place on the premise of the company. Coordination among the responders has been demonstrated and documented, and communication mechanisms have been established to assist in the coordination of not only the emergency responders but also stakeholders (i.e., regional counties and cities) that are concerned about the emergency and impacts to their community. In a planned emergency, the hazards have been previously identified and actions needed to control the hazards are understood and known, along with all the responders have been trained to understand their roles and responsibilities. In particular, the behaviors and impacts of physical, chemical, and biological hazards are fully known and mitigated.

Expectations for performance have been identified and evaluated, and emergency response actions have been focused to achieve a successful outcome to a defined end point. With planned emergency scenarios, all responders, management, and governing agencies (including regional and local communities) understand their roles and responsibilities, and communication mechanisms and expectations for performance are well defined. In the case of unplanned emergencies, some information is known, and some is not.

An unplanned emergency and associated response are based on the premise that either the event was not conceived as being possible, or the magnitude of a planned emergency response was larger than anticipated. In the case of 9/11, it was both. No one had planned, from an emergency response perspective, that a coordinated effort would occur with multiple planes being deliberately crashed into buildings or land; nor had anyone planned to respond to the large number of people killed or injured. Prior to 9/11, emergency responders had prepared to respond to a single event, but not multiple events with large casualties and injuries. As a result of 9/11, the world of

TABLE 9.1
Differences in Types of Emergency Response Actions

Planned Actions	Unplanned Actions
Disaster scenarios are known.	The disaster scenario itself, or the magnitude of the disaster, had not been considered and planned for.
Required notifications have been identified.	Required notifications, including the number and type of notifications, are not fully defined.
Roles and responsibilities are defined.	Roles and responsibilities are defined over time, and initially are based on experience, knowledge, and training.
Resources have been defined and trained.	Resource identification and acquisition are based on availability. Resources include both traditionally thought of emergency response actions and volunteers.
Communication mechanisms have been defined and understood.	Communication is based on the equipment and means available.
Response actions are defined and successfully demonstrated.	Response actions are generally divided between rescue and recovery using available resources and funding.
Response equipment has been acquired and stored.	Equipment is gathered based on its availability and the focus of the initial response actions. Additional equipment is gathered as actions become planned.
Hazards and risks are known and understood.	Initial hazards are known but continue to emerge over time; new hazards can emerge over time.
Hazard controls are clearly defined and executed in sequence.	Hazard controls are based on the identification of initial hazards and then revised as additional hazards are recognized.
The end point has been previously established.	The end point of the response action changes depending on how the rescue effort proceeds. The end point may not be readily recognized if the disaster is significant.

emergency response has evolved and continues to mature to handle new challenges that companies and our country face in protecting workers and residents.

Table 9.1 identifies some of the more distinct differences between planned and unplanned actions by responders. Each of these actions must be considered and decisions made as a part of response execution.

Whether the response actions are planned or unplanned, response to an emergency is implemented in four basic steps: there is an initiating event, event notification, event response, and event or site transition. As relevant to the discussion above, implementation of each of these steps can be planned or unplanned. Figure 9.1 depicts the four steps of an emergency and the associated response for each step is discussed in. How these steps progress, and the decision making associated with several of them, impacts the ability of the emergency response organization to effectively mitigate and protect personnel and the environment. Performance of these steps impact the ability of the emergency response organization to effectively mitigate and protect personnel and the environment.

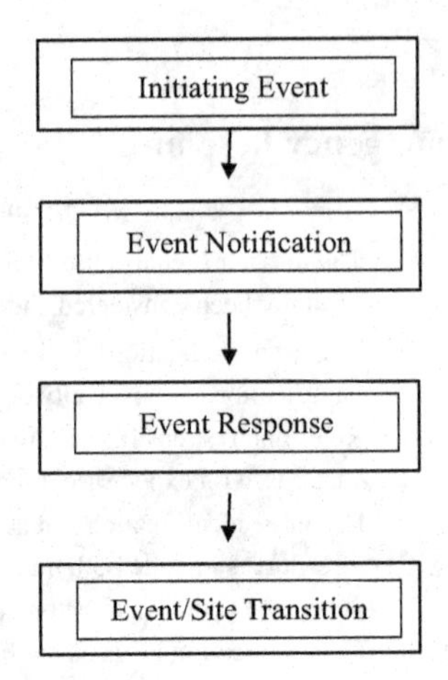

FIGURE 9.1 Four steps of an emergency response.

9.3 INITIATING EVENT

An emergency always starts with an initiating event. Generally, there are two types of threats and disasters that require some type of emergency response:

- Initiating events that occur naturally
- Initiating events that are man-made

Naturally occurring disasters that require an emergency response include fires, hurricanes, flooding, earthquakes, and lightning strikes (to name just a few). Insurance companies refer to these types of disasters as "acts of God." Generally, with emergencies that are initiated by nature, some level of emergency response has been planned, but what may vary is the magnitude of the event (such as Hurricane Katrina). In addition, when responding to natural disasters, more than one location may be impacted by the initiating event, which will drive an increase in the level of coordinated response and communication.

For example, in the case of Katrina, the emergency response effort needed was more significant than originally envisioned or planned because of the level of devastation and the large number of people impacted. Although emergency response organizations may have prepared for the disaster to occur, the impact or secondary effects of the event and the response actions needed to mitigate these impacts were not as well planned because of the size of the response effort needed.

Man-made emergencies are quickly becoming a growing threat, and awareness of the impact of these disasters to people, animals, and the environment continues to increase. Man-made emergencies including traditional workplace accidents, the effects of war, and the threat of terrorism are just a few examples of where a strong emergency response organization is needed. Over the years, society has become increasingly aware of the impact of man-made emergencies and has matured in its level of preparation when responding to such emergencies. In the United States, under the Superfund Amendments and Reauthorization Act (SARA) Title III, also known as the Emergency Planning and Community Right-to-Know Act (EPCRA), companies are required to provide state and local agencies information concerning potential chemical hazards present in their communities, to support emergency planning and response actions. Additionally, 29 CFR 1910.119, Process Safety Management of Highly Hazardous Chemicals, was promulgated to drive companies to plan, analyze, and control exposure to chemicals that can be harmful, and potentially fatal, to workers, the public, and the environment. These regulations were designed to protect workers, the public, and the environment from workplace accidents that occur and to assist in planning emergency response actions, such as the chemical release that occurred in 1984 in Bhopal, India. In recent years, acts of terrorism have become a predominant factor in emergency response planning to man-made emergencies. Because of 9/11, response actions have significantly matured and communication among federal and state agencies has improved.

9.4 EVENT NOTIFICATION

Once an emergency has occurred, notification of the event to some authoritative entity is the start of the emergency response effort. Traditionally, event notification occurs via the telephone or pager. However, in the age of electronics, event notification can occur through the internet or text. Event notification generally occurs to a primary notification recipient; a person who is usually designated as the point of contact for initial notification of an emergency such as a supervisor or control room operator. Generally, the event notification process includes:

- Notifying all affected personnel of an emergency, including corporate entities
- Performing an initial qualitative analysis of risks and conditions
- Evaluating personnel and logistics needs

In the case of a natural disaster, there is usually a warning that the event is going to occur, so the notification process can be communicated through normal channels, such as radio, television, or the internet. The notification process can occur over time, and there is generally preplanning associated with the response effort. In the case of a workplace accident, the notification process is implemented by a triggering event and the notification chain has typically been preestablished and documented in company procedures. However, in the case of a terrorist attack, there may be no notification process other than the event itself and notification of its occurrence.

As part of the event notification process, the person notified usually performs an initial qualitative analysis. Of primary consideration is keeping the response team safe and uninjured. Items to consider when qualitatively evaluating information communicated as part of the initial event notification are listed below:

- Establishment of command and control: Because more than one person will be responding to the event, it is important to establish the hierarchy of authority and the person or persons who will ultimately be held responsible for leading the emergency response effort on the ground. In a planned response, the establishment of command and control is predetermined; roles and responsibilities have been clearly defined. In the case of an unplanned emergency, it is important to define, early in the response effort, who will be making the decisions on how the response effort will be implemented. Traditionally, the fire department, specifically the fire marshal, has served in this capacity because of their significant experience in dealing with emergencies daily.
- Determination of the primary hazard posed by the event and associated risks to response personnel: An analysis of the risks posed to response personnel is initially done to support planning for the on-site response. The primary hazard posed by an event or emergency is usually the most obvious. For naturally occurring emergencies, the primary hazard is typically the event itself, such as water, wind, or fire and the resulting damage. In the case of workplace accidents, it is typically conditions in the workplace as initially observed, such as a physical or chemical hazard. For terrorist attacks, the initial analysis is based on physical conditions and chemical inventories known at the attack site, along with the believed cause of the event.
- Determination of secondary hazards caused by the event and associated risks to response personnel: Again, an analysis must be performed in order to better protect and prepare response personnel. For naturally occurring emergencies, secondary hazards may include destroyed property and roads, transportation, and shelter, needed medical attention, and continued rescue of personnel. In the case of man-made emergencies, secondary hazards posed by the event may include lack of food and water, but also combustion products from isolated fires and physical, chemical, and biological hazards that may initially occur or occur over time. Often, for man-made disasters, which are unplanned, there is limited information regarding secondary hazards other than what was readily observed.
- Status of personnel and the need to rescue people: Initial analysis of an emergency event should include understanding whether members of the public or workers are in need of being rescued. If people need rescue, then priorities for stabilizing the event site and the type of personnel needed to respond to the emergency need to be broader than when responding to the event itself, and the prioritization of response efforts may be driven by the need to save lives.
- Conditions at the event site: When determining who should respond to an emergency, site conditions at the event site need to be generally understood because they could influence the type of people needed to respond. For example, do you know if there is any property damage, and do you have a good place for establishing a centralized command and control? Is the event

site itself safe? What tools or instruments should we take with us when responding? These questions are all related to understanding any barriers posed by the event site itself.

- Logistical needs: When initially notified of an event, a significant consideration is how to keep response personnel safe. Maintaining the safety of response personnel is critical in order to rescue people, but also to stabilize the event site. Another consideration is the availability of electrical power, in particular, because of the dependence of society on the use of technology in our daily work activities. Having electrical power readily available is crucial to effective communications, collection of data, responding to medical emergencies, keeping instrumentation running, and general use as part of the response effort. Additional logistical needs that should be considered include water and food for both rescued personnel and those responding: the disciplines of the personnel responding, such as medical, engineering, military, and health and safety; equipment, such as medical supplies and monitoring instrumentation; needed transportation; and where to establish a base working location. When responding to an event, the location that is established as a "base camp" should be safe and easily obtainable by personnel.
- Communication needs and the ability to effectively communicate: In today's society, almost everyone relies on smart technology to enhance both their professional and personal life. The use of technology in communicating the occurrence of an event, warning others of the event, and responding to the event is extremely important. Because there may be times when the ability to communicate is impacted, in many cases in real time, the need to overcommunicate becomes critical. Through the use of technology, emergency response efforts can be monitored and observed as they are occurring, and the ability to communicate with response personnel, and the decision-making process that response personnel must perform, can be assisted and enhanced. Response personnel can directly communicate with personnel in command and control, and the decision-making process associated with emergency response efforts has been greatly improved through the use of technology. The ability to communicate more effectively helps to assist in improving emergency response efforts, but it can also influence perceptions on how the emergency response effort is progressing. Through the use of technology, communications of events can influence whether the public believes the response agency is providing adequate resources and is responsible both morally and fiscally. It is important to note that the larger the emergency response effort, the greater the amount of communication that is needed to effectively manage both emergency response personnel and the public.
- Special needs: Depending on the event, there may be special needs required for response personnel. For example, some regulations in the United States require medical physicals of response personnel. There may be special equipment or special skills needed to respond to the emergency or

contaminants of concern. Special equipment may be needed if responding to a biological, chemical, or radiological event. Depending on the emergency, response personnel should consider whether there is any specialized equipment needed or requirements that need to be addressed when responding to and stabilizing the event scene.

9.5 EVENT RESPONSE

The event response itself pertains to actions needed, at the event scene, to rescue people and stabilize the emergency. On a daily basis, there are several factors that must be considered to influence actions taken by emergency responders, which include risk prioritization, resources and equipment, logistical support, and communications. Figure 9.2 depicts these factors, and each is further described below.

9.5.1 RISK PRIORITIZATION

Risk prioritization is a process used by personnel to identify and prioritize the management of personnel and resources. In the case of emergency response, risk prioritization relates to how hazards, discovered as part of the response effort, are managed and mitigated. The risk prioritization process uses a hierarchical approach and is independent of the size of the event. Risk prioritization considers risk level and time. The greater the need to mitigate the hazard, the sooner the mitigation must occur. Figure 9.3 depicts an example risk prioritization process as part of an emergency response effort.

The risk prioritization process is used when responding to hazards identified, as part of event response actions, and takes into account not only the primary hazard, or cause of the event, but also secondary hazards that have resulted from the event.

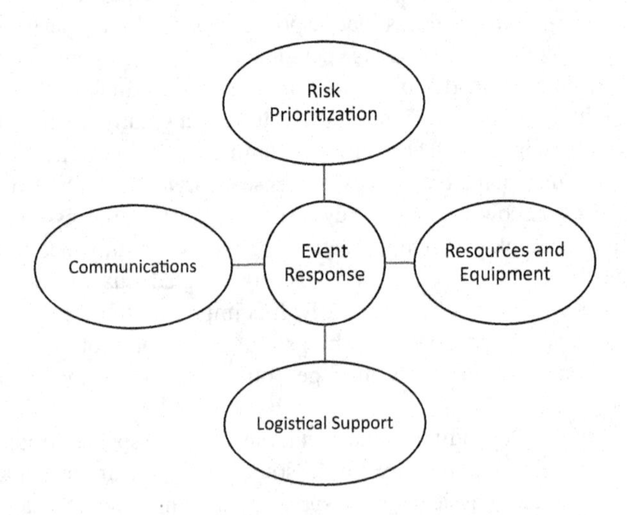

FIGURE 9.2 Event response factors.

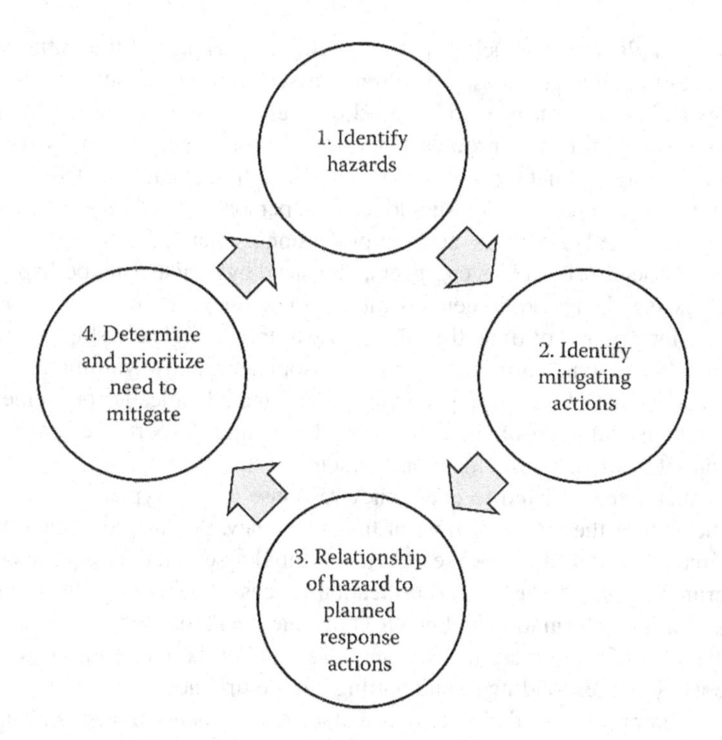

FIGURE 9.3 Risk prioritization process.

When the event notification occurs, there is an initial assessment performed to identify the scope of emergency response actions. Upon arriving at the event scene, a more detailed and in-depth assessment of the hazards is performed. An analysis of the event scene hazards should occur and may cause a reprioritization of management of the hazards. For example, a chemical accident may occur in the workplace, but upon arriving at the scene, electrical or water hazards may also be present, which would be a more imminent threat than first envisioned as a part of the initial emergency response effort.

The risk mitigation process is usually managed by the event response lead and should include feedback and input from the response team. Through application of a risk prioritization, resources can be focused and response to the event, especially emergencies that are large in terms of response effort needed, can be significantly improved, along with providing a safer environment for emergency response personnel to work in.

9.5.2 Resources and Equipment

When responding to an emergency, the primary focus is rescuing personnel and stabilizing the site, preventing additional damage to personnel and the environment. The number and types of people and equipment needed to respond should be identified and planned as much as possible – whether as part of the initial response assessment or as time progresses. The initial assessment of the number and type of resources and equipment needed is based on preliminary information obtained as part of the event notification process. Once at the event scene, a person will get a better understanding of what types of resources and equipment are needed.

The type of resources needed should relate to personnel needs, such as search and rescue, but should also take into account professional disciplines that may be needed for the response effort, such as engineers, industrial hygienists, and perhaps demolition experts (e.g., heavy equipment operators). In addition, if the event response is to occur for a long period of time, then the ability to transition leadership and response personnel also becomes a priority because resources are always finite. At a minimum, a core group of response personnel should include a leader or incident commander, a logistical point of contact, a hazard evaluation expert, medical personnel (if appropriate), and a communications person.

Equipment needs related to emergency response efforts typically increase over time. The shorter the length of time of the emergency, the less equipment that will be required as part of the response effort. An initial assessment of equipment needs is performed as part of the event notification process, but as response efforts continue, and more information is known about the event site environmental conditions, the type of equipment needed may change. For example, initial responders to a fire would be responding to and putting out the fire; however, monitoring at the event location may be needed to ensure that air quality is not impacted. The ability to continue to have access to needed equipment and personnel becomes extremely important as the length of the response action increases.

9.5.3 Logistical Support

Logistical support relates to the procurement and distribution of supplies and resources. If the response effort will be longer than a day or two, then the need to organize and establish processes for obtaining supplies, and the ability to distribute those supplies, becomes a growing concern and one that can cripple the continued response. Although this is a function that can be managed by the event response lead, it is generally delegated and one person is assigned responsibility to ensure that all resources, supplies, and pathways for delivery are unimpaired.

As with any emergency response role, when designating a person to provide logistical support, that should be their sole job – to ensure that all needed supplies are obtained. The designated logistical support person will be managing the needs of not only response personnel in the field, but also those in the command center, as well as arranging for corporate and outside stakeholder meeting locations. This person also works closely with the communications specialist. Considerations when providing logistical support include:

- Availability of electricity, restrooms, water, and food
- Establishment and maintenance of a command-and-control headquarters and secondary response stations

- Establishment and sustainability of communication mechanisms, such as cell phones and radios, computers, and internet services
- Transportation of response personnel
- Ability to refurbish needed equipment and supplies
- Personal protective equipment and ability to dispose of waste

This list is not all-inclusive but serves only as an example of items to consider when providing logistical support. Considerations will vary depending on the emergency, but the role of providing logistical support is critical to being able to sustain the response effort.

9.5.4 Event Response Communication

As the emergency progresses, the need to effectively communicate the status of the event, if there are any injuries, and the current state of the event location becomes an integral part of the response effort. The heart of communication pertains to management of people and outside perceptions of how the response effort is progressing. There are three primary needs of communicating during an emergency:

- Communications within the response team itself
- Communications with stakeholders
- Communication with the public and media

Communications within the response team itself are generally focused on making sure the response team is safe and managing their response actions as they are progressing. If communications are not well established, then it becomes more challenging to ensure that event responders are safe and the decisions being made, as part of the response action, are appropriate.

Historically communications with the response team were conducted by hardwired telephone and at intervals, such as every 30 minutes. In today's electronic environment, there are multiple means available to communicate between the response team and the event response lead. Many corporate and government entities have improved communications with the response team and now are able to communicate in real time as the response team is implementing protective actions. Tools that are useful in establishing effective communication with event responders include cell phones, computers, and portable cameras that the response team lead, along with other interested parties, can use as the emergency response is unfolding. There are also portable electronic devices available that can be set up, both at the event site and at the event response headquarters, whereby planning can be conducted and shared in real time among the response team members and the response team lead.

Communications with outside stakeholders are also needed to ensure that all affected parties are informed of the state of the emergency. Whether it is a federal or state agency, regulatory body, or corporate interest, all these stakeholders are wanting to understand the status of the event so that they may understand their liability and appropriately respond to public inquiries. In the case of the 9/11 event response, regulatory agencies were a part of the planning and mitigation of the event.

Communication with the public and social media has become one of the most impactful needs when responding to an emergency event. The ability of the event response lead and stakeholders to effectively communicate with the public and social media is extremely influential in how the response effort is perceived and whether it is deemed successful. It is important to note that depending on the size of the event, the level of importance of communication with the public and social media grows exponentially. In particular, many people have smartphones and are continually linked to the internet; therefore, social media has become a primary means of communication. If communications are not effective, this can result in an increase in the level of regulatory oversight and funding, and a negative perception regarding how a company treats its employees and cares for the community. Below are some recommendations, currently used by companies today, to improve communications:

- Designate two communication positions or specialists at the start of an emergency: a communications representative at the event scene and a communications representative to communicate with outside stakeholders and regulators, the public, and social media.
- Establish clear roles and responsibilities of event response personnel regarding communication with outside stakeholders and regulatory agencies.
- Establish a routine frequency by which communications are presented to the public and social media.
- Be proactive in communications; however, the communication should be structured and always have a purpose.

When responding to emergencies, it is recommended that a person be designated as a communications representative whose primary job is to coordinate communications to entities outside of the response team itself.

9.6 EVENT OR SITE TRANSITION

Once the initial response effort has been completed, and the initiating event has been mitigated, the emergency response effort is transitioned to another individual who will be designated with returning the site back to pre-event conditions. The event or site transition generally occurs after all personnel have been rescued or obtained medical attention, the initial hazard has been mitigated, any secondary hazards have been identified and are being successfully managed, and a long-term corporate entity has been brought in to manage any long-term site cleanup.

In the case of natural disasters, there is generally a state or city government agency that takes control of areas that were struck. For man-made accidents, the transitional body that will preside over long-term cleanup is generally a company, but it may also be a state or city agency (in the case of an act of terror).

When transitioning personnel from a recovery and response effort, a primary consideration is understanding the end state of when the site response and transition effort is complete. Defining the desired end state of the response action is extremely critical to ensure that the appropriate personnel are identified to support long-term stabilization activities, as well as understanding when deployed resources can be reassigned. When transitioning personnel from the initial response personnel to a governing body that will oversee long-term cleanup and event stabilization, the ability to fund the effort starts to be a primary concern. Initial estimates of cleanup are generally not inclusive of all efforts needed to return an event site to its initial state. Depending on whether the public was harmed, special funds may need to be established to assist families in remediating their home. In the case of Hurricane Katrina, the government set up a special fund to assist families who did not have the ability to pay for the damage to their homes.

Resources needed as part of site transition, management and accountability, funding, and logistical support are all considerations of site transition. In addition, the role of communications will be reduced; however, continuing to effectively communicate continues to be a high priority.

9.7 LESSONS LEARNED FROM 9/11

Probably the most significant emergency response effort in the history of the United States was 9/11. The industrial hygienist played a role in this response effort, and the remainder of this chapter is written to capture lessons learned from an industrial hygienist who was a part of the response effort and was at Ground Zero within a week of the planes striking the WTC.

9.7.1 Event Notification

The event notification for 9/11 was through commercial television, the internet, and radio. The initiating events continued for several hours. Companies that had personnel in the WTC were frantically trying to determine employee accountability – who was in the WTC on that fateful day. Simultaneously, as the event was unfolding companies started realizing that they would be responding to the event to provide assistance and to better understand what response and recovery actions were being conducted at Ground Zero. Consequently, personnel and volunteers were quickly identified and dispatched to assist in the emergency response effort.

Initial response equipment was minimal due to the site being controlled by federal, state, and local agencies. Corporate and institutional responders were not sure what was going to be needed because the effort to rescue people was a primary focus of the response effort, and information regarding an initial risk assessment of what hazards may exist at the event scene was minimal. Instrumentation tailored to the hazards, as they became known, was obtained after additional hazard information became available, along with better respiratory protection. Federal, state, and local municipalities and corporations started identifying resource needs and funding mechanisms for the emergency response effort. Contracts were established among different stakeholders, including large construction companies, to assist in the recovery and response effort.

Medical physicals were not required for response personnel, other than physicals required by environmental and occupational safety regulations. As time progressed a psychological exam was performed for all response personnel prior to working the emergency scene; this was recognized as a good practice because of the number of individual stress factors associated with the response effort (e.g., recovering deceased bodies, having to interact with relatives of missing people). A medical representative individually spoke with each response person to perform general psychological testing and to prepare personnel for responding to the devastating event scene.

9.7.2 Event Response

Response personnel were not sure of the extent of hazards at the event scene; little information was initially available because of the time constraint to rescue people. Response personnel were not sure if there were hazards other than falling debris and building material. Specifically, response personnel did not know if the terrorists had intentionally brought chemical, biological, or radiological contaminants onto the planes that had crashed.

The initial site visit was overwhelming. No one had ever envisioned the magnitude of the devastation, both physically and emotionally, that they encountered when they first arrived at Ground Zero. Response personnel did not know what contaminants were present, other than the event scene had a pungent odor and small combustion fires were everywhere (fires and thermal releases). As time progressed, response personnel determined that exposure to asbestos was limited, and by-products of combustion or subsequent chemical reactions became a significant hazard of concern.

Physical hazards at the site were unpredictable because of the ever-changing conditions as part of the rescue and cleanup effort. The initial focus of the response effort was on physical hazards, addressing acute health hazards that may be present. As the presence of chemicals became known, industrial hygienists recognized that there would be chronic health hazards associated with the response effort. When investigating, response personnel realized that there were a significant number of chemicals that could be present but might not have known which chemical was causing an odor. Most of the odor was coming from underground, originating from under the ground surface, and may have combined with other contaminants.

A risk prioritization process was used in the response to 9/11. New hazards were identified on a daily basis. Because of the size of the response effort, along with the limited number of resources available to assist in the response effort, new hazards were continuously arising and needing to be managed. A risk prioritization process was established and used daily to prioritize risks to the response team and to assist in site stabilization activities. Identified hazards were prioritized on a daily basis, during a morning meeting, to determine the priority in which a response was needed: mitigation immediately needed, mitigation could occur the next day, or mitigation was needed within the week or month. Of primary concern was the need to keep the response team safe from harm when responding to the 9/11 event scene.

Response personnel were assigned to work on rotating shifts. Transportation had to be arranged to deliver personnel to the event scene on a rotational shift basis. Specialized disciplines were used as part of the response effort. Demolition experts, in addition to other traditional response personnel, became part of the response team because they understood hazards that were associated with demolishing a building and how to work more efficiently in areas that contain many physical hazards. The demolition experts provided experience in safe methods to remove large pieces of building debris.

Physical and mental stamina was challenging and difficult. The event scene was a large area, and because of the significant number of people who were injured or deceased, focusing on the job became at times challenging, in particular if someone was looking for their loved one. About a month after the event occurred, golf carts were procured to assist traveling at the event scene; however, riding in the carts themselves also became a hazard because there were no formal roadways.

The primary role of the industrial hygienist was exposure and air monitoring. A large number of air monitoring tests were performed, which included a process for inventorying the samples and controlling and managing data. Because of the changing hazards and conditions, the industrial hygienist had to be flexible and adapt to taking samples under less than pristine conditions. Often, the industrial hygienist was asked to respond to the unknown situation or odor.

A large number of samples were collected and required extensive data entry and management. The need to manage the information originating from the sample collection became challenging because it was difficult to correlate samples previously taken with new hazards that needed to be managed – the data set was extremely large. In hindsight, it was recognized that having one person for data management and analysis would have been useful to better understand and coordinate the overall hazard management process as part of the response effort. People were collecting a lot of data, and the resulting data was entered into a large data collection system, making it difficult for response personnel to understand and effectively manage the hazard management process.

9.7.3 Site Transition

The response and recovery effort for 9/11 went on for several months and years. Because of the magnitude of the emergency, defining an end point to perform site transition was challenging. This is often a major consideration that should be considered as early in the response effort as possible. All agencies and personnel who responded to the event have learned from the emergency. The true extent of health hazards posed to those who responded has evolved from acute to chronic health issues and will continue to impact all those associated with the response. Additional information on the United States' response to the 9/11 can be found at the 9–11 Commission's Report.[4] It is of special note and significance that this chapter is devoted to those who responded to and are continuing to assist in the recovery efforts associated with 9/11 and other terrorist attacks.

Questions to Ponder for Learning

1. What are the two types of threats or disasters that require some type of emergency response? How can the two types be distinguished?
2. What role does the industrial hygienist play in emergency response?
3. Is it feasible to prepare for all types of emergencies? Explain and justify your response.
4. List and explain factors important to an industrial hygienist when responding to an emergency.
5. Discuss the importance of effective communications during emergencies.
6. How have the past emergencies impacted emergency response efforts today?

REFERENCES

1. LaPorte Facility Chemical Safety Board Investigation Report. https://www.csb.gov/dupont-la-porte-facility-toxic-chemical-release-/
2. Collapse of the Rana Plaza. https://www.cnn.com/style/article/rana-plaza-garment-worker-rights-accord/index.html
3. Glasgow Explosion. http://news.bbc.co.uk/1/shared/bsp/hi/pdfs/16_07_09_icl_inquiry.pdf
4. Commission on 9/11. https://9-11commission.gov/report/

10 Evaluating the Industrial Hygiene Program

10.1 INTRODUCTION

A successful industrial hygiene program is built on the fundamental principle that both the program and field execution are being implemented in accordance with processes and procedures that are in compliance with both company and regulatory requirements. As an industrial hygienist, you are frequently required to act as a safety representative in the workplace. Industrial hygienists may collect air samples in the morning, participate in a pre-job midmorning, and observe workers from a safety perspective while work is being performed in the afternoon. Depending on the need of the moment, the workforce often expects the industrial hygienist to be *the* safety person.

Recognizing, evaluating, and mitigating hazards are the fundamental building blocks to protecting the worker, but to ensure that the industrial hygiene program is sound, established on an adequate technical basis, and supports risk reduction in the workplace, the programs must be evaluated to confirm that they are maintained and compliant with laws and regulations, and consistent with the company vision and guiding principles. The primary method to evaluate the industrial hygiene program is by conducting assessments.

The term *assessment* is the act of evaluating and drawing conclusions about the state of a process, program, activity, or event, the act of evaluating something, an idea or opinion about a topic. Assessments can be conducted in many ways:

- Field walk-arounds
- Inspections
- Surveillance
- Independent evaluations
- Documented formal assessments

Depending on the needs of the company or client, assessments can be structured in a number of ways and implemented either formally or informally. ISO 45001, *Occupational Health and Safety Management Systems*, discusses assessments in Chapter 9, *Performance Evaluation*. Assessment types include self-assessments or audits, assessment of management topics (management assessments), and assessments or audits performed by accrediting or independent agencies/personnel.

There are many reasons why an industrial hygiene program should be evaluated. One of the most common drivers for evaluating an industrial hygiene program is to

DOI: 10.1201/9781032645902-10

ensure that the company is compliant with worker safety and health laws and regulations – which is linked to worker protection and company liability. Listed below are a few examples of why a program or process should be assessed:

- To understand whether the program and field implementation is compliant with laws and regulations – minimize regulatory, personal, and financial risk
- At the beginning of assuming control of an industrial hygiene program – establish program baseline
- Periodically to ensure the program is still compliant with laws and regulations – this information is useful when being investigated
- To fulfill the requirements of insurance companies that regular evaluations are conducted to ensure the adequacy of the program's content
- To understand the level of regulatory, financial, and human health risk that exists or may have changed because of an event that occurs
- Upon receipt of an enforcement action, to identify areas that need to be improved to minimize future risk
- Based on complaints by workers, to understand and reduce human health and regulatory risk

Typically, assessments and other means for evaluating programs and processes address three fundamental areas:

- Compliance to requirements
- Performance of workers, processes, and programs
- Quality of executing and performing tasks or processes

The reasons for performing an assessment may vary, but the workflow when conducting an assessment is generally the same. Figure 10.1 depicts the general workflow of conducting an assessment. Each element of the workflow process is presented in this chapter, along with tools that are useful to the industrial hygienist in planning, conducting, and using assessment results to both further improve programs and processes used by a company and confirm that they ultimately support a reduction in the overall risk.

10.2 IDENTIFYING THE PROGRAM AND PROCESS TO ASSESS

Assessments generally target a program or process that needs to be evaluated to determine if it is functioning as designed. The program or process may be at the company level, such as evaluating whether the program or process used in reporting safety statistics across the company or corporation is being performed in accordance with company procedures. Assessments may also be conducted at the facility level, such as determining whether the facility public address system is providing effective notification of emergencies to building personnel. Facility assessments target specific organizations, divisions, or locations (for example). The most common use

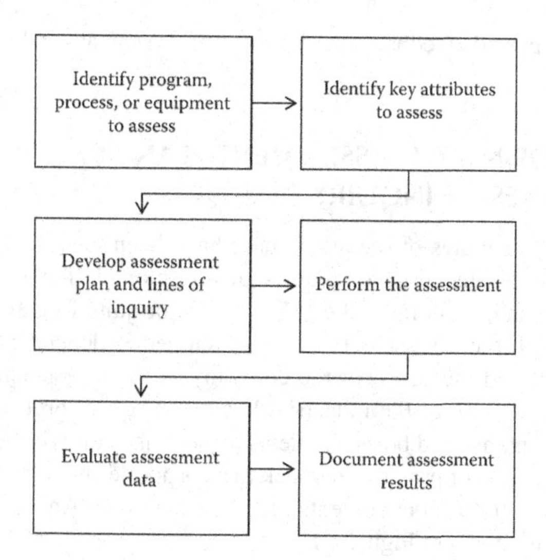

FIGURE 10.1 Workflow for conducting an assessment.

of assessments is to evaluate worker or field performance at the activity level. A good example of an activity level assessment is observing a work pre-job and watching the work being performed in the field. No matter what the topic is being assessed, the workflow for performing an assessment remains the same.

10.3 IDENTIFYING KEY ATTRIBUTES TO ASSESS

Key attributes refer to those characteristics or traits of a program or process that make it unique or vital to ensuring that the program or process functions as designed. For example, as an industrial hygienist or safety professional, you may be asked to evaluate whether the company's exposure assessment process is being implemented as intended. Key attributes that would be targeted for being assessed could include:

- Evaluating whether an exposure assessment strategy has been documented and is being implemented
- Ensuring the adequacy of workplace data collection
- Establishing accurate exposure profiles and similar exposure groups
- Identifying and implementing hazard controls based on exposure profiles

Identification of the correct critical attributes is important because evaluation of the attributes will drive the focus and value of the assessment. For example, to determine whether the safety culture of a company is healthy, critical attributes to be assessed would include criteria reflective of a healthy safety culture, as defined by the assessor, company, or standard.

The assessor should also consider who the customer is for the assessment; the customer may have specific feedback as to the attributes to be evaluated, or the worker may also provide feedback on specific attributes to assess based on their field experience.

10.4 DEVELOPING THE ASSESSMENT PLAN AND LINES OF INQUIRY

Once the critical attributes of the assessment have been identified, then an assessment plan and associated lines of inquiry are developed. The development of an assessment plan can be tailored to be either simple or quite formal and lengthy, and the extent to which the assessment plan is developed is dependent on the purpose of the assessment and the needs of the company. If the assessment is to be a self-assessment/internal audit of the industrial hygiene program group, then the format of the assessment plan could be as simple as a checklist. If it will be used to support ongoing legal litigation or provide feedback to a corporate entity on areas to improve, then more elaborate criteria documents may be warranted. An example of a simple assessment plan is shown in Figure 10.2.

Company X **Assessment plan**
Assessment topic: Implementation of the asbestos program
Person performing assessment: Susan B. Anthony
Assessment timeframe: July 21, 2017–August 6, 2017
Assessment methods: Walk-through of the area. Review of program documentation. Interviewing two workers.
Assessment lines-of-inquiry: 1. Does the company maintain a list and location of asbestos contaminated material? Select several work locations to confirm that all required signs and postings are visible. 2. Have workers been provided initial and continuing training as required? 3. Are workers provided the appropriate PPE commensurate with program requirements? 4. Are the results of sampling work evolutions below regulatory and company requirements? 5. Is a written program in place, which is compliant with regulatory requirements, and requires all appropriate monitoring? 6. Are appropriate respirators used when required? 7. Are employees who perform work with asbestos in a medical surveillance program? 8. Do workers feel they are protected? 9. What opportunities exist for improving program and field implementation? 10. Does the company have an effective asbestos program?

FIGURE 10.2 Assessment plan template.

A typical assessment plan should include the following:

- Assessment topic
- Person performing the assessment
- Assessment time frame
- Assessment methods
- Lines of inquiry to be used when conducting the assessment (which could consist of a ready-made checklist)
- Assessment results

If the assessment is being informally conducted, then a written plan is not necessary; however, depending on how the information from the assessment will be used, it is always recommended to document the assessment methods and results in some manner for future reference, whether it be in a logbook or an informal database. The information can be used for communicating to management or workers that a work process has been evaluated and determined to either be adequate or needs improvement. Assessments performed for the industrial hygiene program are also useful during litigation or regulatory investigations because it demonstrates the company is responsible.

As part of the assessment plan, the method of assessment will need to be defined. Assessment methods include the following:

- Reviewing program or process documentation
- Interviewing workers and management
- Observing work in the field

As an assessor, you will decide the best method for performing the assessment based on the topic and needs of the assessment. One or more assessment methods are generally used in any type of assessment. The assessment methods are used to implement the lines of inquiry.

Lines of inquiry are typically written in the assessment plan as a question, but could be in the form of a statement, and can be used with any assessment method (e.g., whether you are interviewing a worker in the field or reviewing documentation that is used to validate safety and health information communicated to the workforce). Lines of inquiry typically originate from program and/or process procedures and form the fundamental basis for performing the assessment.

The reason why the assessment is being performed will drive how the lines of inquiry are written. If the assessment is being conducted to determine whether the program or process is compliant with regulatory requirements (compliance-based), then the lines of inquiry will be developed from the requirements in the form of a question or a statement. For example, in the United States 29 CFR 1910.95(c) requires employers to administer a continuing, effective hearing conservation program whenever employee noise exposures equal or exceed an 8-hour time-weighted average (TWA) sound level of 85 dB on the A scale or, equivalently, a dose of 50%. A line of inquiry that can be used to evaluate this requirement may consist of,

Does the company require an employee to be in a hearing conservation program if the employee noise exposures equal or exceed an 8-hour time-weighted average (TWA) sound level of 85 dB on the A scale or, equivalently, a dose of 50%?

The method used to evaluate this line of inquiry could include performing and/or reviewing noise monitoring results and then evaluate whether the employees have been enrolled or need to be enrolled in a hearing conservation program. When assessing risk, it is important the lines of inquiry reflect the type of risk being evaluated such as financial risk, health risk, company risk, and how the risk is impactful.

The number of lines of inquiry developed will vary depending on whether the scope of the assessment is large or small. There is no defined number of lines of inquiry needed; however, the number of lines of inquiry should be sufficient to draw conclusions of the assessment that support whether the program or process is adequate or if improvement is needed. Additional factors to consider when developing lines of inquiry include:

- What information is being sought when reviewing documentation?
- What are the work practices that the workers will be following when performing the task to be observed?
- How will the information from the assessment be used, and are there other considerations, such as legal or personnel requirements, which should be incorporated into the lines of inquiry?
- How can one communicate the questions in such a manner that both the people being assessed and the personnel who will use the information understand?

10.5 PERFORMING THE ASSESSMENT

Performing the assessment may seem obvious; however, there are several items that should be considered when performing and implementing an assessment activity. First and foremost is the need for the assessor to communicate the assessment prior to performing the activity. Most employees and managers become nervous when someone comes to their work site and communicates that he or she will be performing an evaluation. The natural tendency of people is to be concerned that they will do something wrong and get in trouble. It is important that the assessor communicate the following, preferably several days in advance:

- The purpose of the assessment
- How the assessment will be conducted
- That no blame or discipline will result from evaluating the work
- That the outcome of the assessment will be communicated to management and the employees
- That feedback as to how the assessment can be better performed will be solicited
- That any questions that may arise prior to the assessment being performed will be answered

Results of the assessment will be better accepted and appreciated if the assessor effectively communicates the purpose and scope of the assessment and enlists the support of management and the workforce prior to initiating the assessment. To assist the industrial hygiene professional, there are general rules to be considered when assessing.

- Try and put the workers who are being assessed at ease. Open the visit with a good explanation of how the assessment will be conducted, what scope will be assessed, and how the information will be used. Often, a joke or communicating that workers' concerns are understood, prior to the assessment being initiated, is a good idea.
- Make sure to communicate what your role is and that you are there to help. Communicating and helping the workers understand that everyone is on the same team, and that you yourself are a worker, assists in building trust.
- Communicate that assessments are a normal part of doing business. If the company or function does not understand how well they are performing against requirements, then improvement cannot be achieved. This is also a good opportunity to receive feedback and tailor your lines of inquiry toward worker concerns or areas in which they would like to understand how the process or program is functioning.
- There is no blame when assessing. The assessor is not here "to get you"; they are here to help.
- Solicit trust and collect feedback from the workers and incorporate that information into the lines of inquiry.
- When interviewing, find a location that is neutral and not threatening. If the workers would like to bring someone with them as a witness, that should be embraced.
- Communication is key to a successful outcome. If anyone has any questions, they should be encouraged to raise them, and together the assessor and assessed personnel can resolve them.
- The assessors should not outnumber the workers being assessed. Often, during interviews assessors will have two or three people interviewing one person. Such an environment can create apprehension and fear that the interviewee is being set up. In some cases, they may refuse to be interviewed because they feel uncomfortable.
- When observing work, if an unsafe condition exists, then it should be communicated to the supervisor or worker performing the unsafe act. At times, it may be appropriate to stop the work. The assessor will lose credibility and trust if an issue related to an unsafe working condition is raised after the fact. The worker will then believe the assessor is "out to get them."
- A good assessor looks for every opportunity to build trust and respect during the assessment process. In particular, the industrial hygienist is considered a safety and health representative and, as such, is expected to have integrity and character. After all, the workers are putting their safety into the industrial hygienist's hands.

As with every task the industrial hygienist performs, sufficient documentation and justification as to how the assessment was performed and the results are important. Because many tasks that an industrial hygienist performs support technical decisions, and possibly future litigation; thorough and adequate documentation is a necessity.

10.6 DATA ANALYSIS

Once the field work is complete for the assessment, information and data generated from the assessment must be analyzed and conclusions drawn. Data from the assessment will vary depending on whether it is qualitative or quantitative; an industrial hygienist is required to have the skills to evaluate both types of data.

Qualitative data is more often associated with program reviews and are typically performance-based. For example, is the industrial hygiene program or process conforming to the company procedure or regulatory requirements? Are personnel adequately performing, implementing, and reporting in accordance with company requirements? Is the program or process effective while being implemented? How well has the identified risk been mitigated?

Quantitative data is more often associated with process reviews and mathematical data. For example, when reviewing hearing data, does it meet or exceed the regulatory threshold for enrollment in a hearing conservation program? Are personnel appropriately enrolled in a medical monitoring program if sampling data indicates that workers are exposed to chemicals or contaminants above a certain threshold level? In many cases, both qualitative and quantitative data are evaluated as part of assessing industrial hygiene practices. Once the data has been generated, it must be organized, analyzed, related to the existing programs and processes, and stored. Figure 10.3 depicts the relationship of all these factors in the analysis of data.

10.6.1 Data Organization

The manner by which the data is organized is important because the data itself may be used in a number of ways. Listed below are some of the more common reasons why data must be organized and should be considered when organizing data generated from assessments.

- The data must be organized to support conclusions of the assessment. Every assessment has a purpose, and how the data is organized will support conclusions drawn to meet the need for why the assessment was conducted. Whether the assessment was conducted to determine program compliance to regulations or to determine whether a program is being effectively executed, how the data is organized should support and provide an answer to why the assessment was conducted.
- Conclusions of the assessment must be reproducible and credible. How the data is organized can either assist in achieving reproducibility and credibility or hinder and draw criticism from others reviewing the assessment.

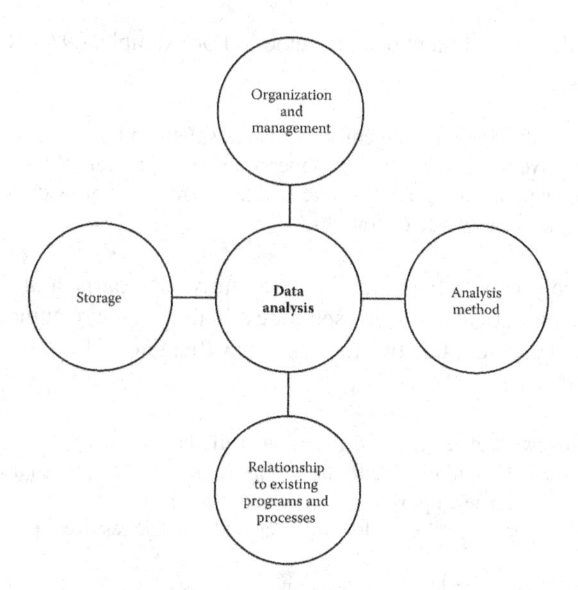

FIGURE 10.3 Elements of data analysis.

Having the data organized in such a fashion that lends credibility to the assessment will be of significant importance to the industrial hygienist for future reference. There is nothing worse than an outside party reviewing assessment data and not being able to draw the same conclusions from the data. This is of particular importance if the assessment will be used, or possibly relied on, in regulatory litigation.

- The data must be organized in a logical fashion to effectively support the evaluation and assessment conclusions. If the industrial hygiene professional cannot show logically how the data was organized, then credibility becomes an issue, and even if the data is accurate, the illogical manner in which it is organized will not positively drive others to believe that it is valid.
- The assessor is writing a story on how they performed the assessment and final conclusions. The assessor should be writing the assessment from their eyes and explain how the data supports the story and overall conclusions.

10.6.2 Analysis Method

The method by which the data is analyzed is extremely important because it can drive the defensibility of conclusions drawn from the assessment. Data may be analyzed either qualitatively or quantitatively, and in both cases, data is compared against criteria and a cognitive analysis is performed.

Qualitative analysis of the data is based on a cognitive comparison of the data against criteria that are typically nonnumerical. Criteria used for evaluating data are generally based on a statement or requirement. For example, 29 CFR 1910.120(b)(1)(i) requires that

> employers shall develop and implement a written safety and health program for their employees involved in hazardous waste operations. The program shall be designed to identify, evaluate, and control safety and health hazards, and provide for emergency response for hazardous waste operations.

Quantitative analysis of data is based on a cognitive comparison of the data against criteria that are numerical. Criteria used for evaluating data are numerical, and often analysis of the data uses statistics to verify or validate conclusions. Appendix F of 29 CFR 1910.95 states:

> In determining whether a standard threshold shift has occurred, allowance may be made for the contribution of aging to the change in hearing level by adjusting the most recent audiogram. If the employer chooses to adjust the audiogram, the employer shall follow the procedure described below for each audiometric test frequency:
>
> - Determine from Tables F-1 and F-2 the age correction values for the employee by:
> - Finding the age at which the most recent audiogram was taken and recording the corresponding values of age corrections at 1000 Hz through 6000 Hz;
> - Finding the age at which the baseline audiogram was taken and recording the corresponding values of age corrections at 1000 Hz through 6000 Hz;
> - Subtract the values found in step (i)(B) from the value in step (i)(A).
> - The differences calculated in step (ii) represented that portion of the change in hearing that may be due to aging.

This book will not go into detail as to how statistics are used in the industrial hygiene process; however, what is important is that the industrial hygienist understands how statistics can be used to strengthen the quality of the data and support data analysis conclusions. In most cases, the industrial hygienist will employ both qualitative and quantitative analysis in making a determination as to whether a regulation has been exceeded. Results from a personal lapel sampler may, at first glance, indicate that chemical concentrations were above regulatory limits; however, because the worker was not present in the workplace for 8 hours, further analysis determined that chemical concentrations did not exceed the worker protection value and poses a minimal health risk.

10.6.3 Risk Management of Identified Hazards

A key consideration in the analysis of data gathered as part of an assessment is understanding the relationship between the identified condition and associated risk. The risk may be financial, health-related, perceived, regulatory, or environmental.

There are all types of deficiencies associated with identified risk, and it is up to the industrial hygiene professional to identify, understand, and determine the best method for risk management associated with protecting workers.

When conducting assessments, risk is generally identified during the phase of comparison of the data to qualitative or quantitative criteria. Risk is also prioritized from greatest to least risk consequences. The identified risk can also be cumulative, meaning that as additional risk is identified, the overall consequences of risk increase. The acceptability of risk to the workers, company, and industry can also influence the analysis of data and formulated conclusions. Additionally, the cultural impact of accepting or reducing risks identified during assessments must be considered.

In the age of the internet, society has become less accepting of risk, which to the industrial hygiene professional equates to increased performance in managing and controlling hazards in the workplace. It is recommended that when evaluating data generated from assessments, the industrial hygienist has a good understanding of the risk threshold of the company, and regulatory agency, when controlling exposures and hazards to workers in the workplace. One such example is that the acceptability of the risk and hazard thresholds, for protecting workers when conducting remediation work under the Resource, Conservation, and Recovery Act (RCRA) is different from that under the Occupational Safety and Health Administration (OSHA). In the United States, OSHA typically requires harm to have been proven such as workplace injury or illness; however, under RCRA, there only needs to exist the potential for injury or illness before additional personal protective actions may be warranted.

10.6.4 Relationship of Data to Existing Programs and Processes

Data generated as a part of the industrial hygiene program almost always contributes, to some degree, to other functional elements of the overall industrial hygiene program or processes, along with those programs and processes associated with other company functions. For example, data generated as part of a routine sampling program may be used in the work control process, such as defining the level of engineering controls or personal protective equipment needed to safely perform work and adequately protect workers. The data may also be used in understanding whether workplace conditions have changed over time, or determining whether daily work is being safely performed, or possibly whether operational equipment is degrading. Routine industrial hygiene data is also extremely valuable when issues occur in the workplace and validating personnel were, and have not been, acutely or chronically exposed to chemicals or particulates. It is important that the industrial hygiene professional understand where and how the data that is being generated, as part of routine sampling, can be used in other programs and processes that the company uses in everyday work.

10.6.5 Data Storage and Management

Data storage and management is not necessarily a topic that many industrial hygienists consider; however, it is becoming increasingly important in the current technology age, along with society becoming less tolerant to work environments that can cause serious health effects. There are several regulations that require exposure records to be maintained for a specified period of time. One of the most common regulations is 29 CFR 1910.1020(d)(1)(ii), which states that "each employee exposure record shall be preserved and maintained for at least 30 years." Depending on the industry or function of a company, there may be additional requirements associated with data retention and storage (such as records). When storing and managing industrial hygiene data, the industrial hygiene professional should consider the following:

- Data retrievability: Industrial hygiene data that is generated must be stored to enable retrievability for an indefinite amount of time. This can be challenging because most data generated in the twenty-first century is electronically generated and stored. Programs that are currently being used for storing and retrieving data may not be available in the future – and most likely will not be. So, it is up to the industrial hygiene professional to monitor and update data management software on some routine frequency. Even if the data is saved on a storage device, such as a memory drive, in 5–10 years the memory drive may not be readily retrievable. It is recommended that the industrial hygiene professional consult with computer professionals when determining the best way to store and retrieve exposure and other industrial hygiene data.
- Quality management standards: Most companies are managed under some type of a quality assurance program. For many years, there has not been a good integration of quality assurance standards and industrial hygiene data, or even the implementation of an industrial hygiene program. Unless an industrial hygiene program has established a quality assurance program that is as stringent as the company quality assurance requirements, it is very likely that all aspects of the industrial hygiene program may be regulated, to some extent, by quality assurance. Depending on the industry or company in which the industrial hygienist is employed, there may be quality assurance standards or criteria that must be complied with when storing and managing industrial hygiene data and records. It is a good idea that the industrial hygiene professional understands the quality standards that regulate his or her company or industry, and how those standards may apply to their program including data storage and management.
- Ease of accessibility: Many computer programs used by industrial hygiene professionals have some form of requirements or controls associated with accessing data. The industrial hygiene professional needs to understand any limitations or controls that may be present, such as read and/or write capability, and whether there are security features that should be enabled when using and controlling accessibility to the data. Shortcuts can be created that can ease accessibility, and it is important to understand whether more than one person, at the same time, can access the data. This is extremely important because if more than one person cannot be using the data at the same time, then often data is not entered and saved in the program (when

the industrial hygienist thought it was saved). Also, security passwords are helpful to ensure that only those personnel authorized to use the data have access.

- Custodial responsibilities: All companies experience some amount of employee turnover. It is common in today's work environment for industrial hygiene professionals to transition, or progress, to another job with increased responsibilities. Often, this progression does not happen within the same company long term. It is important that one person, typically a supervisor or manager, maintain custodial ownership over any databases associated with industrial hygiene exposure monitoring or data. The manager or supervisor may not always be the database administrator (that responsibility can be delegated), but overall custodial responsibility should reside with a person of authority, and one that is most likely to stay with the company. Many companies today use change management plans to manage key company processes, and storage and management of industrial hygiene data, because of legal liabilities, should be designated a key process. As personnel transition to other jobs, either inside or outside the company, the change management plan is used to identify company risks and how to mitigate and manage them, such as maintaining custodial responsibility of the industrial hygiene data.
- Interfaces with occupational medical programs and providers: Industrial hygiene data is used in a variety of ways, one of which is that information is used in occupational medical programs to determine whether an injury or illness is work related. The interface between an industrial hygiene program and its designated occupational medical provider is important, because without the data the provider cannot make an appropriate medical determination. The industrial hygiene professional should understand the electronic database interfaces between the two programs and/or companies and ensure that data is routinely and accurately transmitted to the occupational medical provider.

10.7 DOCUMENT ASSESSMENT RESULTS

Once the data has been analyzed, conclusions are formulated, and information is then documented in some form of an assessment report. It is always a good practice for the assessor to meet with the employees or management who were involved in the assessment, prior to formally documenting and issuing the assessment, to ensure that everyone understands the conclusions of the assessment and to gain their acceptance of the assessment findings. This meeting also provides employees and management an opportunity to give feedback on whether they agree with the conclusions, or if there is information that was overlooked by the assessor that might impact the assessment conclusions.

Documentation of the assessment can vary, from a small, one-page checklist with a summary at the end to a multipage report with conclusions and recommendations identified. The assessor must understand how the information from the assessment may be used, which will help determine the level and depth of how the assessment is documented. The understanding of risk associated with the information generated from the assessment should be included in the assessment report, and then communicated to all employees involved in the assessment. This is extremely important in building trust and respect with the workforce. A typical assessment report consists of the following:

- Assessment topic
- Assessor name and assessment time period
- Approach and methodology
- Assessment criteria
- Data results
- Conclusions and recommendations

The assessment report should also define how the assessment information will be used and how the assessment report itself will be managed.

Many assessment reports recommend a path forward for resolving any issues identified during the assessment as a way to assist the organization or group being assessed. Depending on the industry or company, there may also be quality assurance requirements that drive how the assessment is documented. The industrial hygienist should have a good understanding as to whether the company or industry has standards associated with documenting assessments.

Questions to Ponder for Learning

1. List some examples of why a program or process should be assessed.
2. Describe the workflow process for conducting an assessment.
3. What are the components of a typical assessment plan?
4. When functioning as an assessor, what are some of the general rules that you must consider?
5. Discuss the elements of a data analysis and the input on the assessment process.
6. Discuss the importance of conducting an assessment that is viewed as credible and reproducible.
7. List some of the many ways that risk can be defined.
8. What attributes of data storage must be considered when storing collected data that represents workplace hazards and exposure evaluations?
9. What are the component contents of a typical assessment report?

11 Continuous Improvement

11.1 INTRODUCTION

The industrial hygiene profession is one that is focused on identifying and controlling workplace hazards, including physical hazards and associated risks, by ensuring the safest work environment for the employee. Both the industrial hygiene program and implementation of the program in the field can always be improved – and should be improved – to ensure compliance with the most current requirements, standards, and safety practices, and to effectively manage risk.

The term *continuous improvement* means specific actions, or a series of actions, that the industrial hygienist or company executes to improve safety and health or industrial hygiene programs with respect to productivity, compliance, or actions to minimize health risks to employees. The International Organization for Standardization (ISO) 45001 standard, *Occupational Health and Safety Management Systems*, Section 10, *Improvement*, is focused on continuous improvement of the safety and health organization, which includes industrial hygiene.

Continuous improvement of the industrial hygiene program targets functional elements of the program that have been identified as not being well implemented or have been fully implemented but a higher level of performance is desired. For example, over the past years several self-assessments have identified noncompliances in the hearing conservation program. Two employees were identified as performing work in an area where noise greater than 85 decibels was present, but neither employee was enrolled in the company's hearing conservation program. Another self-assessment identified industrial hygienists were developing exposure assessments for three maintenance work packages; however, monitoring results from the jobs were not consistent with the industrial hygiene analysis and calculations. In both examples, several actions could be taken to improve the overall industrial hygiene program and company performance in either the noise control program or the hazard identification and control process.

The ultimate goal of continuously improving the industrial hygiene organization is to reduce overall risk to the company and its employees. The risk reduction could be legal, such as improving compliance to a specific regulation. The risk reduction could be financial, such as reducing the cost to implement the industrial safety and hygiene program because of automated monitoring equipment. Or, the risk reduction could be to human health, such as designing a ventilation system to operate below 85 decibels. Whatever the reason, continuous improvement in industrial hygiene should

DOI: 10.1201/9781032645902-11

improve current functional operations, reduce liability, as well as improve the work culture and conditions in the field.

11.2 CONTINUOUS IMPROVEMENT PROCESS

Compliance to the ISO 45001 standard requires development of industrial hygiene processes that drive continuous improvement. Specifically, Section 10 of ISO 45001 requires the organization to consider results from analysis and evaluation of occupational health and safety performance, evaluation of compliance, internal audits, and management reviews when taking action to improve. Figure 11.1 depicts a continuous improvement process which can be applied to the industrial hygiene organization and program to further reduce operational and financial risks.

A continuous improvement program for industrial hygiene should include:

- Benchmarking: Benchmarking consists of comparing an industrial hygiene program and processes with the programs and processes associated with companies that are recognized as strong performers. The goal in performing benchmarking is to raise expectations of existing program performance to a higher and better standard that will result in improving protection of the workforce. When performing benchmarking, the industrial hygienist identifies those companies or institutions that are recognized as the "gold standard" and evaluates their program against the recognized "gold standard" programs, identifying those areas of their program where improvements could be best realized. Because the majority of industrial hygienists are employed by a company or institution (not generally self-employed), the

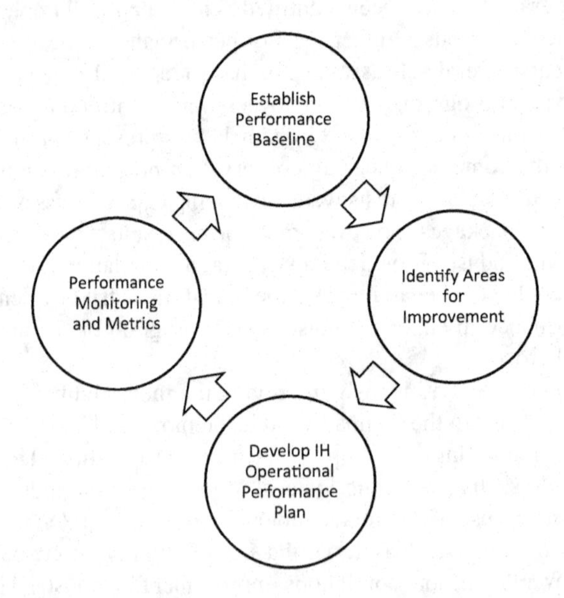

FIGURE 11.1 Example industrial hygiene continuous improvement process.

cost and time of implementing such an improvement must be considered and the actions must be beneficial to all parties (such as senior management and workers) involved in supporting the improvement plan.

- Management and professional presence in the field: Improving the performance of industrial hygiene processes and procedures does not need to just be focused on the industrial hygiene or safety and health discipline; continuous improvement initiatives should also target cultural aspects of the safety program. One tool used in improving overall safety and health performance is increasing the level of visibility of management, supervisors, and professionals in the field and on the shop floor. Depending on the culture of the company, often an effective industrial hygiene program is improved through increasing the level of trust that exists between union and non-union workforces. Improving the relationship between management and the workforce will result in greater ownership of the industrial hygiene program by everyone in the company, because it has become personal to them. It is also of importance that scheduled management visits be planned to further build a positive work environment and establish routine communication forums for the workforce.
- Performance metrics: Performance metrics are used to evaluate how well implementation in the field is meeting program and performance expectations. For example, the company may have a goal to evaluate compliance with two subparts of 29 CFR 1910 within a 12-month period. The performance metric would establish a baseline of two evaluations per month, but because the company has had trouble keeping equipment operating due to maintenance issues, competing priorities, and scheduled personal vacation, only one evaluation was performed over a six-month period. The performance metric would indicate that either additional resources are needed to achieve adequate performance of the industrial hygiene function, or the priorities of work being performed need to be evaluated. It is important that the senior management of the company or institution accept and support the performance metrics, because they are generally the individuals in authority who can take actions to achieve and improve performance. They should also be driving improved performance of all functions within the organization. In addition, educating the workforce on these performance metrics will assist in achieving the goals.
- Employee engagement and involvement: Having the employees engaged and involved in improving safety performance has always been a crucial element of a successful industrial hygiene program. The workers often have the best ideas for how to address the hazard and improve safety performance. Building employee engagement and involvement into the industrial hygiene program and field implementation starts with establishing and maintaining personal relationships with the workforce. Involving them in implementation of the industrial hygiene program also educates them and improves their understanding of the role of the industrial hygienist and empowers them to self-police their work environment.

11.3 ESTABLISHING A PERFORMANCE BASELINE

Continuous improvement begins with defining the current performance baseline of the industrial hygiene program. This task includes establishing and defining how industrial hygiene processes and procedures are implemented in the field in comparison with a similar company whose program has been recognized for excellence – benchmarking.

When defining the performance baseline, it should be evaluated over a period of time to thoroughly understand how the work is truly conducted in the field. In addition, the evaluation should identify what tasks were performed, by whom, and the process or procedure steps that were performed.

Several continuous improvement tools can be used in defining the performance baseline. Assessment of the program is a fundamental tool and was described in greater detail in Chapter 10. Use of a series of assessments, coupled with benchmarking those program attributes, where improved performance is desired, against programs considered the gold standard in industry, can become a powerful tool from which to understand the current basis and state of the industrial hygiene and worker protection programs.

When determining the performance baseline, it is important to remember that the baseline is not being established to criticize how the program or process is being implemented, but rather to identify the level of performance that currently exists, so that improvement in performance can be identified and executed over time.

11.4 IDENTIFY AREAS FOR IMPROVEMENT

Over time, incidents occur, and weaknesses are identified in field implementation of the industrial hygiene program. Non-compliances and risk management issues are typically documented, need to be evaluated, and appropriately managed. Under the ISO 45001 standard, incidents and nonconformances (i.e., procedure non-compliances) are required to be investigated and reviews to be conducted to ensure corrective actions are appropriate and address causes of the underlying issue.

It is important to understand that under the ISO 45001 standard a nonconformance is defined by a non-fulfillment of a requirement, and a requirement is defined as a need or expectation that is stated, generally implied, or obligatory. This is often confusing because a procedure noncompliance can by definition be a nonconformance to the program.

The ISO 45001 standard identifies example incidents and nonconformities which should warrant further evaluation:

- Incidents such as fall with or without injury, broken leg, asbestosis, hearing loss, and damage to buildings or vehicles where they can lead to occupation health and safety risk.
- Nonconformities such as protective equipment not properly functioning, failure to fulfill legal requirements and contractual requirements, and prescribed procedures not being followed.

Chapter 10 identifies a number of methods which can be used to evaluate both program and field performance, but once the data is collected it is required to be evaluated. Evaluation of data can either be qualitatively or quantitatively performed. Data that is numerical in nature is typically quantitatively evaluated against a numerical standard or regulation (e.g., Threshold Limit Values, Permissible Exposure Limit). In addition, when the industrial hygienist is evaluating performance data (qualitative data), the data is typically compared against a goal or benchmark. The cause of the incident or nonconformance is evaluated and corrective actions identified which could mitigate the incident in the future.

As part of the data evaluation process opportunities for improvement should be identified, along with corrective actions associated with a particular incident or nonconformance. Often, activities and steps needed to complete the corrective actions are documented in a continuous improvement plan for industrial hygiene.

Many companies have quality assurance standards that must be met. These companies have a corrective action management process, a subset of the quality assurance program, which may define how to evaluate data, performance cause analysis, and development of corrective actions. Often corrective actions are organized by significance or risk to be mitigated; higher risk issues are given a higher priority in resolution vs. lower risk issues; as well as, identification of the root vs. apparent cause of the event. Incidents with similar causes can be grouped and managed for collective significance and evaluation. When identifying corrective actions and areas for improvement, the industrial hygienist should consider actions in terms of being specific, measurable, achievable, relevant, and timebound (SMART):

- The corrective action or improvement initiative should be specific. For example, the organization will perform assessments using the company procedure and demonstrate implementation of each procedure step.
- The corrective action should be measurable. For example, the organization will increase the number of industrial hygiene assessments from one per year to six per year.
- The corrective action or improvement initiative should be attainable. Often organizations identify actions which cannot be performed because either the completed corrective action is not realistic, or technology or people are not (or would not be) available to perform the work.
- The corrective action or improvement initiative should be relevant to the issue or improvement initiative. For example, the company will increase the number of industrial hygiene personnel who are qualified to perform assessments (which will make it easier to meet the increased requirement number of assessments).
- The corrective action or improvement initiative should be timebound. Completion of actions should be reasonable and not allowed to be completed over a long period of time. The industrial hygienist will also want to identify whether the corrective action and improvement initiative is compensatory in nature (short term because the significance of the issue is low) or long term and designed to prevent recurrence.

The identification of corrective actions and improvement initiatives should involve the workers, to instill ownership and understand effectiveness. The workers are one of the greatest assets to the industrial hygienist in understanding how to improve compliance and performance, but also teaming in further reducing operational risk.

There are a number of resources for the industrial hygienist to use when identifying corrective actions and improvement initiatives. Feedback solicited from peers and work crews is an excellent source of improvement initiatives, along with reviewing lessons learned from the previous year. Many federal agencies in the United States have guidance on improvement programs, along with professional societies, and information can also be obtained through internet searches. In addition, professional colleagues in similar industries can be a good resource for ideas.

11.5 INDUSTRIAL HYGIENE CONTINUOUS IMPROVEMENT PLAN

An Industrial Hygiene Continuous Improvement Plan is written when an increased level of performance is needed in the industrial hygiene function. Larger companies or companies that do work in standard based industries (i.e., nuclear) encourage the development of improvement plans so they can demonstrate compliance to the requirements, but also demonstrate how risk is reduced to their corporate mission. It should also be recognized that in the United States, there is no federal or state regulation that requires an industrial hygienist to establish and implement a continuous improvement program; it is based on the commitment of the company, the safety and health organization, management, and workers to improve how they are performing work and striving to provide a safer work environment.

If the industrial hygiene program conforms to the ISO 45001 Occupational Health and Safety (OH&S) standard, the industrial hygienist must have a program that analyses and manages program performance including compliance, self-assessments, and management reviews. The industrial hygiene organization is required to perform incident investigations, reporting of noncompliances and nonconformance, and corrective action management. In addition, ISO 45001 specifically identifies examples of continual improvement issues, which may include, but not limited to:

- New technology
- Good practices, both internal and external to the organization
- Suggestions and recommendations from interested parties
- New knowledge and understanding of occupational health and safety-related issues
- New or improvement materials
- Changes in worker capabilities or competence
- Achieving improved performance with fewer resources (i.e., simplification, streamlining, etc.).

The industrial hygienist should develop a continuous improvement plan that is based on the continuous improvement process and use tools that will facilitate continuous improvement. The continuous improvement plan should identify the vision of the organization or group, include organizational goals and objectives, identify corrective actions and/or improvement initiatives, and identify a method for measuring performance. The plan should be written for a specified time period and should identify not only programmatic benefits to be achieved, but also benefits to the company so that management can recognize the benefit of accepting and sponsoring the plan.

Acceptance and sponsorship of the plan by management will not only be of financial benefit, but also gain organizational and company support for improvement of the industrial hygiene or safety and health program. It should be noted that the continuous improvement process is not just limited to the industrial hygiene program but can be applied whenever improved performance is desired. In addition, a continuous improvement plan does not need to be lengthy; documentation of the plan can be as simple as two pages. Of greatest importance is identifying, documenting, and completing improvement actions, measuring progress, and realizing improved performance.

11.5.1 Goals and Objectives

The purpose of developing goals and objectives is to define a level of performance that the industrial hygiene program strives to achieve; often, goals and objectives are annually identified. A goal simply represents the desired performance to be achieved. For example, a goal of the industrial hygiene program may be to have a highly skilled industrial hygiene organization. An objective defines a measurable end point by which to attain or accomplish a goal. Consistent with the example, an objective of the industrial hygiene program would be to have half of the industrial hygiene professionals complete two continuing education classes a year. Both goals and objectives should be revisited on an annual basis, or whenever a management change has occurred. Figure 11.2 presents an example of goals and objectives documented in an industrial hygiene continuous improvement plan.

There are no hard rules when developing goals and objectives; however, a general rule of thumb is to define no more than four to six goals, and then one or two objectives per goal. If there are too many goals and objectives, then they may not achievable because the industrial hygienist and company have limited resources (both people and funding) to complete the plan.

Inherently, people want to do a good job at work and be successful; if there are too many actions to complete, in addition to routine job responsibilities, the staff can become overwhelmed and frustrated. In addition, implementation of goals and objectives is most successful when input from workers and management occurs to facilitate individual ownership and enhance the feeling that they have directly contributed to the success of the organization.

11.5.2 Corrective Actions and Improvement Initiatives

Corrective actions and/or improvement initiatives from the Industrial Hygiene Continuous Improvement Plan should be integrated into daily operations so that improvement becomes a part of the everyday business. Not only does the action

Industrial Hygiene Program Goals for Calendar Year 2018
1. The industrial hygiene staff is highly qualified and recognized for their expertise.
2. State-of-the-art instrumentation is used in workplace monitoring.
3. Risk-based decision making is central to providing a workplace free of hazards.
4. Industrial hygiene staff are actively involved and participate in the work planning process.
5. Workers are effectively trained in the identification and control of hazards.
6. Management and workers own the safety and health program.

Industrial Hygiene Program Objective for Calendar Year 2018
1(a) Provide two continuing training classes each year to both Industrial Hygiene professionals and management, and one leadership class for the Industrial Hygiene Manager.
1(b) A development plan is written for each safety and health professional which contains personal and professional goals.
2(a) Computers used in the Industrial Hygiene Department are upgraded and use the most current data analysis software.
2(b) A third of the instrumentation used in the Industrial Hygiene monitoring program is replaced.
3(a) A quantitative risk analysis process is defined and incorporated into existing procedures and data analysis processes.
3(b) Industrial Hygiene risks have been identified, documented, and being actively managed.
4(a) An Industrial Hygienist is involved in the evaluation of all work planned.
5(a) All workers are trained on the work planning process.
5(b) The Industrial Hygienist is included as part of the emergency response team.
5(c) The process for responding to known emergencies has been identified and institutionalized.
6(a) An employee-owned safety committee is established and regularly meets.
6(b) Performance metrics have been developed to monitor effectiveness of the employee-owned safety committee.

FIGURE 11.2 Example of goals and objectives of a continuous improvement plan.

or initiative have the best opportunity to get completed, funding and additional resources needed to accomplish the work have been minimized. When developing corrective action and improvement initiatives for the improvement plan, the industrial hygienist should consider:

- Personnel and training that may be required.
- Additional funding to support the additional personnel training.
- Working with management to fund the actions or initiative.
- Process used to obtain funding and concurrence with management.

Figure 11.3 presents an example of improvement actions identified to meet specified goals and objectives. Although the example pertains to meeting a goal and objective, the same process applies for managing corrective actions and improvement initiatives.

Industrial Hygiene Continuous Improvement Plan Calendar Year 2018		
Goal: State-of-the-art instrumentation is used for workplace monitoring.		
Objective: 1. Computers used in the Industrial Hygiene Department are upgraded and use the most current data analysis software.		
Action	**Due date**	**Status**
1. Identify computers to be upgraded.	Within one month	Complete
2. Research and identify desired data analysis software, associated cost, training required to operate system.	Within one month	Complete
3. Identify minimum computer capabilities needed to run data analysis software (e.g., amount of memory needed).	Within one month	Complete
4. Develop a cost estimate and submit to management for approval	Within two months	Management approval received
5. Upon receipt of money, place order to procure software. Include in procurement any training necessary from the manufacturer.	Within three months	Ongoing; order will be placed within week
6. Schedule software installation and training classes.	Within six months	On schedule

FIGURE 11.3 Example actions needed to meet a goal or objective.

Actions associated with the improvement plan should be measurable, "statused," and updated monthly. The status of the plan should be reviewed with staff, and feedback solicited, so that adjustments can be made to further assist in achieving goals and objectives within current company schedules. In addition, management may wish to have one person be responsible for ensuring that routine reviews of the plan have been conducted.

Once implementation of the continuous improvement plan has been completed, a summary of the benefits achieved should be communicated to management and the workers, along with ideas for improvement initiatives for the next operational or calendar year. The continued acceptance and sponsorship of these initiatives by management and the workforce will be key in being able to implement future initiatives for continuous improvement. Without management and workforce sponsorship, the ability of the industrial hygienist to acquire the funding and resources needed to continue to implement a sound program will be impacted. Continuous improvement of any program can only be achieved when management, professionals, and the workforce come together to own and manage the industrial hygiene program.

11.6 PERFORMANCE METRICS AND MONITORING

Measuring and evaluating performance should be an inherent component of any industrial hygiene program. Continuous improvement cannot be realized if the plan for improving performance is not measured. Measuring and evaluating performance should utilize and integrate several continuous improvement tools, such as assessments, employee engagement and involvement, and management presence in the field. Several tools can be used together when implementing improvement actions. These tools can be used by themselves, in implementing actions, or they can be used for measuring and evaluating performance.

For example, if an improvement action is to schedule biweekly employee-owned safety committee meetings, then a tool that can be used to measure performance could be a performance metric that identifies the goal of two meetings per month versus actual meetings scheduled. Another example is that improvement is desired in building the relationship between the industrial hygienist and workers; a performance metric could be developed that identifies the number of times per month the industrial hygienist attends worker pre-job briefings. Measuring and understanding feedback from management, industrial hygiene, and worker interactions can prove to be invaluable because the end result of such interactions may result in an increased level of trust and relationship building.

Generally, most companies review their operational performance on a monthly or frequent basis. This type of review provides a good forum for the industrial hygiene manager or staff to summarize and present the continuous improvement plan for acceptance. Performance metrics are the primary means by which to measure improvement. Figures 11.4–11.6 are examples of typical performance metrics used in a continuous improvement program to monitor implementation of a continuous improvement plan. It is important to note that the results obtained from reviewing performance measures should be communicated across the company, and not just within the industrial hygiene and safety and health groups, in order to strengthen

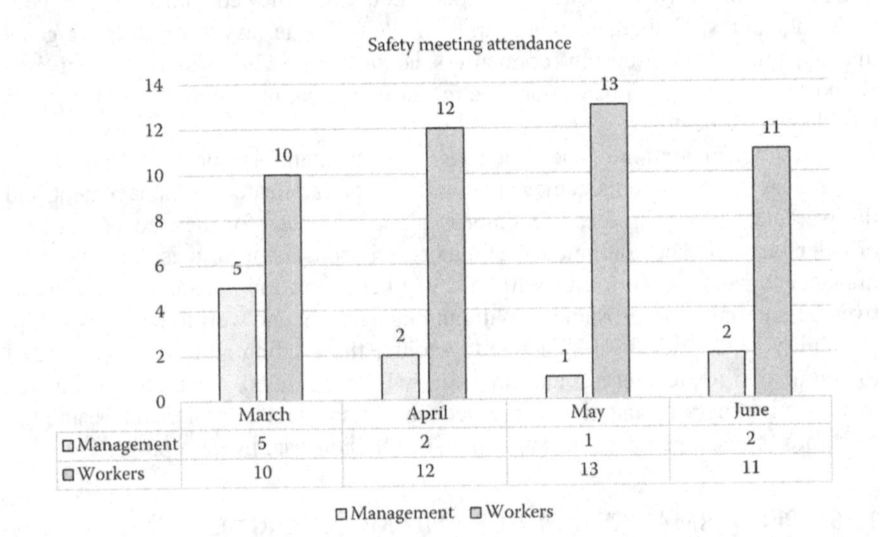

FIGURE 11.4 Example metric of management and worker attendance at employee-owned safety meetings.

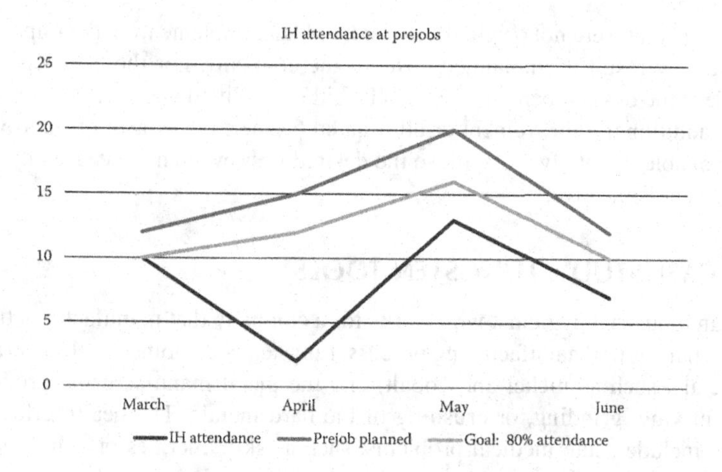

FIGURE 11.5 Example performance metric of industrial hygienist attendance at pre-job meetings.

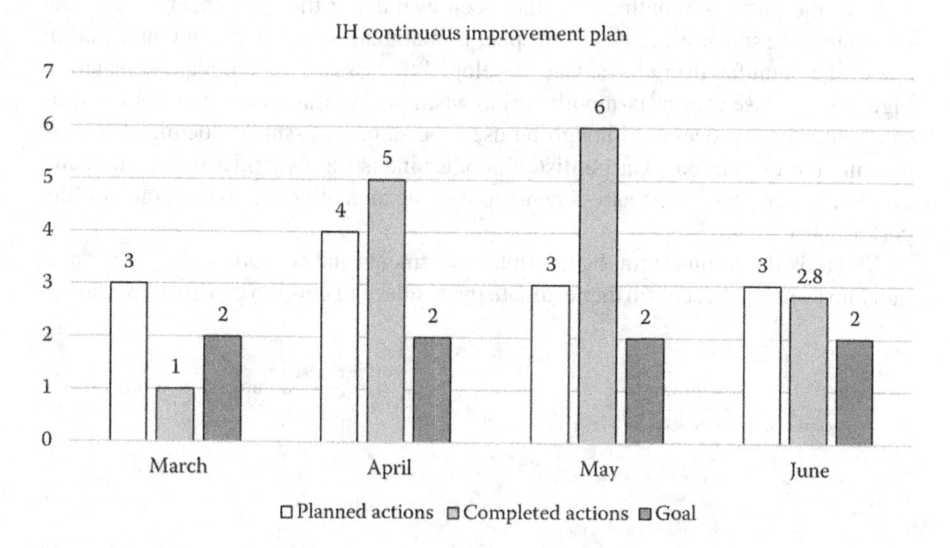

FIGURE 11.6 Metric on planned and completed actions for industrial hygiene (IH) continuous improvement plan.

the importance of the worker protection program. In addition, review of the metrics should be an occurrence that is scheduled as part of routine operations.

When reviewing the metrics or other means to evaluate the achievement of performance improvements, the industrial hygienist needs to recognize those parts of the plan that have achieved the desired performance improvement, and those that have not. Sometimes improvements to the program may actually yield a negative consequence or not achieve the desired benefit at all. In addition, there may be hidden costs that were not originally considered, and implementing the improvement initiative may result in unplanned costs to the company. For those actions that do not achieve the desired benefit, they can be either modified and carried into the next year for additional improvement or eliminated because of the lack of improvement. It is acceptable to not always achieve the desired improvement, because any level of improvement is positive.

11.7 CASE STUDY: TUNGSTEN TOOLS

Carl is an industrial hygienist who works for a company that manufactures tungsten tools. As part of the manufacturing process, tungsten is combined with several other hard metals, such as nickel and cobalt, and the metal manufacturing process can include mixing, grinding, or crushing of the hard metals. The health effects from tungsten include acute medical problems, such as skin allergies or skin burns, and chronic exposure to tungsten can cause lung scarring (pulmonary fibrosis) over time. The company uses containment and negative ventilation as the primary engineering controls to minimize exposure of workers to the metals.

Over the past six months, Carl has been evaluating the implementation of the exposure assessment process with respect to maintenance work being conducted in one of the manufacturing bays. Carl developed the assessment template depicted in Figure 11.7 to use over a six-month period when performing evaluation of the exposure assessment process. Although he used the same assessment document in performing the evaluations, Carl enlisted another industrial hygienist in the company to assist in performing the assessments to provide an additional opinion on baseline performance.

Carl took the results from the multiple assessments and consolidated the information into a spreadsheet to further evaluate the results and develop conclusions. Earlier

Line of inquiry		Company B	Company C
1. Does documentation associated with the work/job identify all contaminants the worker could be exposed to?			
2. What specific contaminants have been identified for the planned work?			
3. Has the Industrial Hygienist been involved in the planning of the work/job in terms of performing an analysis of previously identified air sampling data, representative of the work being conducted?			
4. Are the calculations, or modeling performed, valid and correct and support a protective work environment for the worker?			
5. Have engineering controls been identified to control potentia worker exposures?			
6. Is the exposure assessment documentation readily available for review and easily retrievable?			
7. Have the workers been informed of the hazards associated with work being performed?			

FIGURE 11.7 Exposure assessment evaluation template.

in the year, Carl had attended a professional conference and was able to meet and network with other industrial hygienists who work for tungsten tool manufacturing facilities. In particular, there were two tungsten tool manufacturing companies, Companies A and B, that were known to have strong management sponsorship of their safety and health programs, and their people had been recognized for their efficient and effective programs. Carl contacted both companies and was able to obtain copies of their exposure assessment processes.

Carl identified key attributes of the exposure assessment programs from the other companies and evaluated his programs and processes to identify where there were differences and determine which attributes were worth adopting at his company. The results of the evaluation were documented, and several actions related to improving the exposure assessment process were identified and incorporated into a continuous improvement plan.

QUESTIONS TO PONDER FOR LEARNING

1. Explain why an industrial hygienist would want to develop a continuous improvement plan for his or her company.
2. What are the tools an industrial hygienist can use in developing and implementing an industrial hygiene continuous improvement plan? Explain the significance of each.
3. What are the elements and challenges of an industrial hygiene continuous improvement plan?
4. What benefit is there to the industrial hygienist in defining a performance baseline?
5. How does the industrial hygienist develop performance goals and objectives? Identify several goals and objectives.
6. Define and describe how to measure and evaluate performance.
7. Develop performance metrics related to an industrial hygiene continuous improvement plan.

Index

Accountability, 5, 41, 159
Administrative controls, 8, 59, 61, 99, 102, 134
Administrative Order on Consent (AOC), 93
American Board of Engineering and Technology (ABET), 33
American Industrial Hygiene Association (AIHA), 16, 33, 45, 60
Analytical method, 33, 51, 56
Assessment plan, 166, 167, 176
Authorizing work, 103

Bangladesh, 147
Behavior, 10, 40, 41, 44, 89, 97, 101, 107, 108, 110, 111, 135, 148
Beryllium, 23, 26, 58, 59, 87, 91, 119, 124, 125
Biological hazards, 13, 25, 86, 87, 91, 92, 98, 100, 148, 152

Carcinogen, 54, 58, 102, 128
Case study, 69, 70, 123–125, 138, 143, 188
Characterization, 16, 46, 48, 53–55, 59, 64
Chemical hazards, 13, 25, 87, 98, 117, 151
Chemical inventory, 76, 99
Chemical Safety and Hazard Investigation Board (CSB), 16, 90
College curriculum, 34, 44
Communication for results, 30
Competence, 2, 31, 127, 182
Comprehensive Environmental Response, Compensation, and Liability Act (CERCLA), 5, 93
Concern for employees, 32
Continuing education, 10, 38, 44, 183
Continuous improvement plan, 181–187, 189
Continuous improvement process, 178, 182, 183
Controlling hazards, 11, 15, 16, 101, 111, 173
Corrective actions, 82, 146, 180–184
COVID-19, 80, 87, 100

Data analysis, 20, 46, 170–172, 176
Data collection, 9, 24, 35, 48, 49, 53, 59, 161, 165
Data management, 9, 161, 174
Data organization, 170
Deepwater Horizon, 90
Detector tubes, 64, 65, 67

East Palestine, 92, 93
Employee engagement, 19, 30, 109–111, 135, 179, 186
Energy Employees Occupational Illness Compensation Program (EEOICP), 56
Engineering controls, 8, 21, 32, 34, 45, 61, 102, 134, 173, 188
Environment Safety and Health (ES&H), 34
Ergonomic hazards, 25
Ethical aspects of industrial hygiene, 40
Event notification, 149–151, 155, 156, 159
Event response, 146, 149, 150, 154–158
Expert witness, 33, 38
Exposure assessment, 8, 11, 17, 20, 22, 31, 45–47, 50, 53, 54, 59, 64, 165, 177, 188, 189
Exposure assessment model, 45, 46, 54
Exposure monitoring, 16, 22, 45, 47–50
Exposure monitoring strategy, 16, 22, 24, 50–52, 58, 59, 62–64, 145, 175

Fixed air monitors, 68
Flint Michigan, 4

Generalist, 28, 39, 40
Glasgow, 147
Global harmonization system (GHS), 97

Handheld electronic monitors, 65, 67, 71
Hazard anticipation, 130, 131, 139–142
Hazard elimination, 102, 134
Hazard inventory, 94, 95
Hazard mitigation, 133, 134, 143
Health and Safety Executive (HSE), 1, 115
Hierarchy of hazard control, 81, 101
Humility, 41
Hurricane Katrina, 145, 150, 159

Industrial hygiene certification, 37
Industrial hygiene courses, 34, 35, 44
Industrial hygiene discipline, 3, 11, 68
Industrial hygiene program elements, 7, 8, 11, 79, 80
Industrial hygiene risk, 1, 74, 75
Initiating event, 149, 150, 158, 159
Injury and illness, 3, 14, 17, 20, 23, 95, 96, 113
Instrumentation and calibration, 9, 11, 21, 22, 53, 54, 67, 68, 81, 118
International Organization for Standardization (ISO), 7, 11, 25, 78–82, 126, 127, 129, 133, 163, 177, 178, 180, 182

Job rotation, 38, 39

Key attributes, 83, 165, 189
Kingston Tennessee Valley Authority (TVA), 93

La Porte, Texas, 146
LNT model, 54, 58, 59
As low as reasonably achievable (ALARA), 58, 59, 80

Management by walk-around, 107
Maslow's hierarchy, 3, 4, 12
The Massachusetts Factory Act, 1
Medical monitoring, 113
Medical surveillance, 113
Monitoring equipment, 42, 51, 64, 177
Monitoring plan, 51, 53–55
Monitoring strategy, 14, 16, 22, 24, 46, 50, 64

National Institute of Occupational Safety and Health (NIOSH), 48, 51, 56–58, 60, 61, 67

Occupational exposure limits (OELs), 47, 48, 52, 57–61, 69, 70
Occupational health program, 14, 17, 18, 22–24, 81
Occupational health provider, 17, 18, 20, 22, 23, 115, 121, 122
Occupational Safety and Health Administration (OSHA), 1, 3, 7, 16, 25, 26, 40, 48–50, 56, 59, 60, 78, 90, 91, 97, 100, 110, 114, 117, 122, 123, 127, 129, 173
Openness and honesty, 31
Organizational structure, 7, 27–30

Peer review, 62, 84
Performance metrics, 179, 185, 186
Performing the assessment, 167, 168, 188
Personal air sampling pumps, 64, 67
Personal protective equipment (PPE), 8, 31, 45, 59, 61, 101, 102, 124, 132, 133, 138, 140, 142, 157, 173
Physical hazards, 13, 25, 26, 86, 87, 98, 111, 160, 161, 177
Plan-Do-Check-Act (PDCA), 78, 79
Product substitution, 21, 61, 102
Professional industrial hygienist, 25
Professionalism, 38, 41
Professional judgement, 46
Project closeout, 105–107

Qualification, 10, 82, 117, 119, 134
Qualitative analysis, 151, 171
Qualitative assessment, 50, 71, 75,
Quantitative analysis, 172
Quantitative assessment, 49, 71, 75
Questions to ponder, 11, 24, 43, 71, 84, 111, 125, 143, 162, 176, 189

Radiological hazards, 5, 13, 17, 26, 53, 87, 100, 113
Rana Plaza, 147
Re-evaluation plan, 64
Regulatory inspections, 47, 100
Reliability, 32, 41
Reporting of Injuries, Diseases, and Dangerous Occurred Regulations (RIDDOR), 95
Resource, Conservation, and Recovery Act (RCRA), 5, 173
Retention, 28, 30, 36–38, 44, 128, 137, 142, 174
Risk acceptance, 31, 72, 84, 85
Risk assessment, 16, 20, 31, 35, 48–50, 62, 71, 73–76, 84, 108, 147, 159
Risk assessment elements, 74
Risk based industrial hygiene, 72, 73, 75, 77, 79, 81, 83
Risk communication, 82, 83
Risk prioritization, 154, 155, 160
Risk ranking, 77
Risk register, 76

Safety professionals, 1
Safety through design, 8, 108, 109
Scope of work, 52, 55, 105
Similar exposure groups (SEGs), 17, 51, 56, 57, 117, 118, 120, 122, 124
Site transition, 149, 150, 158, 159, 161
Specific, measurable, attainable, relevant, and timebound (SMART), 181

Task hazard inventory, 94
Trainer knowledge and qualification, 134
Training effectiveness, 135
Training employee, 111, 129
Training methods, 136
Training program, 2, 39, 126, 128, 129, 135
Training strategy, 128, 129, 143
Trustworthiness, 42

United Kingdom (UK), 1, 93
United States Department of Labor, 6

Vapor monitor badges, 66. 67

Williams-Steiger Act, 1, 11
Work planning, 32, 54, 59, 95, 102, 103, 132
World Trade Center (WTC), 146, 159

Printed in the United States
by Baker & Taylor Publisher Services

Printed in the United States
by Baker & Taylor Publisher Services